IN ASSOCIATION WITH
✗SQA

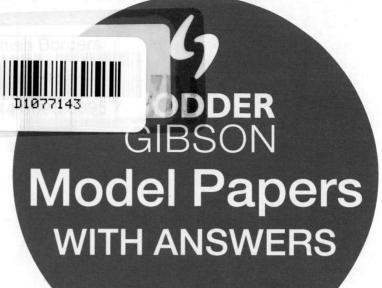

HODDER GIBSON

Model Papers
WITH ANSWERS

PLUS: Official SQA 2014 & 2015
Past Papers With Answers

National 5
Geography

Model Papers, 2014 & 2015 Exams

HODDER
GIBSON
AN HACHETTE UK COMPANY

This book contains the official SQA 2014 and 2015 Exams for National 5 Geography, with associated SQA approved answers modified from the official marking instructions that accompany the paper.

In addition the book contains model papers, together with answers, plus study skills advice. These papers, some of which may include a limited number of previously published SQA questions, have been specially commissioned by Hodder Gibson, and have been written by experienced senior teachers and examiners in line with the new National 5 syllabus and assessment outlines, Spring 2013. This is not SQA material but has been devised to provide further practice for National 5 examinations in 2014 and beyond.

Hodder Gibson is grateful to the copyright holders, as credited on the final page of the Answer Section, for permission to use their material. Every effort has been made to trace the copyright holders and to obtain their permission for the use of copyright material. Hodder Gibson will be happy to receive information allowing us to rectify any error or omission in future editions.

Hachette UK's policy is to use papers that are natural, renewable and recyclable products and made from wood grown in sustainable forests. The logging and manufacturing processes are expected to conform to the environmental regulations of the country of origin.

Orders: please contact Bookpoint Ltd, 130 Park Drive, Milton Park, Abingdon, Oxon OX14 4SE. Telephone: (44) 01235 827720. Fax: (44) 01235 400454. Lines are open 9.00–5.00, Monday to Saturday, with a 24-hour message answering service. Visit our website at www.hoddereducation.co.uk. Hodder Gibson can be contacted direct on: Tel: 0141 848 1609; Fax: 0141 889 6315; email: hoddergibson@hodder.co.uk

This collection first published in 2015 by
Hodder Gibson, an imprint of Hodder Education,
An Hachette UK Company
2a Christie Street
Paisley PA1 1NB

Typeset by Aptara, Inc.

Printed in the UK

A catalogue record for this title is available from the British Library

ISBN: 978-1-4718-6058-4

3 2 1

2016 2015

Introduction

Study Skills – what you need to know to pass exams!

Pause for thought

Many students might skip quickly through a page like this. After all, we all know how to revise. Do you really though?

Think about this:

"IF YOU ALWAYS DO WHAT YOU ALWAYS DO, YOU WILL ALWAYS GET WHAT YOU HAVE ALWAYS GOT."

Do you like the grades you get? Do you want to do better? If you get full marks in your assessment, then that's great! Change nothing! This section is just to help you get that little bit better than you already are.

There are two main parts to the advice on offer here. The first part highlights fairly obvious things but which are also very important. The second part makes suggestions about revision that you might not have thought about but which WILL help you.

Part 1

DOH! It's so obvious but …

Start revising in good time

Don't leave it until the last minute – this will make you panic.

Make a revision timetable that sets out work time AND play time.

Sleep and eat!

Obvious really, and very helpful. Avoid arguments or stressful things too – even games that wind you up. You need to be fit, awake and focused!

Know your place!

Make sure you know exactly **WHEN and WHERE** your exams are.

Know your enemy!

Make sure you know what to expect in the exam.

How is the paper structured?

How much time is there for each question?

What types of question are involved?

Which topics seem to come up time and time again?

Which topics are your strongest and which are your weakest?

Are all topics compulsory or are there choices?

Learn by DOING!

There is no substitute for past papers and practice papers – they are simply essential! Tackling this collection of papers and answers is exactly the right thing to be doing as your exams approach.

Part 2

People learn in different ways. Some like low light, some bright. Some like early morning, some like evening / night. Some prefer warm, some prefer cold. But everyone uses their BRAIN and the brain works when it is active. Passive learning – sitting gazing at notes – is the most INEFFICIENT way to learn anything. Below you will find tips and ideas for making your revision more effective and maybe even more enjoyable. What follows gets your brain active, and active learning works!

Activity 1 – Stop and review

Step 1

When you have done no more than 5 minutes of revision reading STOP!

Step 2

Write a heading in your own words which sums up the topic you have been revising.

Step 3

Write a summary of what you have revised in no more than two sentences. Don't fool yourself by saying, "I know it, but I cannot put it into words". That just means you don't know it well enough. If you cannot write your summary, revise that section again, knowing that you must write a summary at the end of it. Many of you will have notebooks full of blue/black ink writing. Many of the pages will not be especially attractive or memorable so try to liven them up a bit with colour as you are reviewing and rewriting. **This is a great memory aid, and memory is the most important thing.**

Activity 2 – Use technology!

Why should everything be written down? Have you thought about "mental" maps, diagrams, cartoons and colour to help you learn? And rather than write down notes, why not record your revision material?

What about having a text message revision session with friends? Keep in touch with them to find out how and what they are revising and share ideas and questions.

Why not make a video diary where you tell the camera what you are doing, what you think you have learned and what you still have to do? No one has to see or hear it, but the process of having to organise your thoughts in a formal way to explain something is a very important learning practice.

Be sure to make use of electronic files. You could begin to summarise your class notes. Your typing might be slow, but it will get faster and the typed notes will be easier to read than the scribbles in your class notes. Try to add different fonts and colours to make your work stand out. You can easily Google relevant pictures, cartoons and diagrams which you can copy and paste to make your work more attractive and **MEMORABLE**.

Activity 3 – This is it. Do this and you will know lots!

Step 1

In this task you must be very honest with yourself! Find the SQA syllabus for your subject (www.sqa.org.uk). Look at how it is broken down into main topics called MANDATORY knowledge. That means stuff you MUST know.

Step 2

BEFORE you do ANY revision on this topic, write a list of everything that you already know about the subject. It might be quite a long list but you only need to write it once. It shows you all the information that is already in your long-term memory so you know what parts you do not need to revise!

Step 3

Pick a chapter or section from your book or revision notes. Choose a fairly large section or a whole chapter to get the most out of this activity.

With a buddy, use Skype, Facetime, Twitter or any other communication you have, to play the game "If this is the answer, what is the question?". For example, if you are revising Geography and the answer you provide is "meander", your buddy would have to make up a question like "What is the word that describes a feature of a river where it flows slowly and bends often from side to side?".

Make up 10 "answers" based on the content of the chapter or section you are using. Give this to your buddy to solve while you solve theirs.

Step 4

Construct a wordsearch of at least 10 × 10 squares. You can make it as big as you like but keep it realistic. Work together with a group of friends. Many apps allow you to make wordsearch puzzles online. The words and phrases can go in any direction and phrases can be split. Your puzzle must only contain facts linked to the topic you are revising. Your task is to find 10 bits of information to hide in your puzzle, but you must not repeat information that you used in Step 3. DO NOT show where the words are. Fill up empty squares with random letters. Remember to keep a note of where your answers are hidden but do not show your friends. When you have a complete puzzle, exchange it with a friend to solve each other's puzzle.

Step 5

Now make up 10 questions (not "answers" this time) based on the same chapter used in the previous two tasks. Again, you must find NEW information that you have not yet used. Now it's getting hard to find that new information! Again, give your questions to a friend to answer.

Step 6

As you have been doing the puzzles, your brain has been actively searching for new information. Now write a NEW LIST that contains only the new information you have discovered when doing the puzzles. Your new list is the one to look at repeatedly for short bursts over the next few days. Try to remember more and more of it without looking at it. After a few days, you should be able to add words from your second list to your first list as you increase the information in your long-term memory.

FINALLY! Be inspired...

Make a list of different revision ideas and beside each one write **THINGS I HAVE** tried, **THINGS I WILL** try and **THINGS I MIGHT** try. Don't be scared of trying something new.

And remember – "FAIL TO PREPARE AND PREPARE TO FAIL!"

National 5 Geography

The exam

The course assessment will consist of two components: a question paper and an assignment.

The question paper

The purpose of this question paper is to assess your application of skills, and breadth of knowledge and understanding across the three units of the course.

This question paper will give you an opportunity to demonstrate the following higher-order cognitive skills and knowledge and understanding from the mandatory content of the course:

- using a limited range of mapping skills;
- using a limited range of numerical and graphical information;
- giving detailed descriptions and explanations with some analysis.

The question paper will have 60 marks (75% of the total mark) distributed across three sections.

Section 1: Physical Environments

This section will have 20 marks and will be made up of limited/extended-response questions. These require you to draw on your knowledge and understanding, and to apply the skills you have acquired during the course.

Section 2: Human Environments

This section will have 20 marks and will be made up of limited/extended-response questions. These require you to draw on your knowledge and understanding, and to apply the skills you have acquired during the course.

These questions will draw on the knowledge and understanding and skills described in the "further mandatory information on course coverage" section.

Section 3: Global Issues

This section will have 20 marks and will be made up of limited/extended-response questions. These require you to draw on your knowledge and understanding, and to apply the skills you have acquired during the course.

These questions will draw on the knowledge and understanding and skills described in the "further mandatory information on course coverage" section. In this section you will be required to attempt two questions from six. The choice of topics are: Climate Change; Impact of Human Activity on the Natural Environment; Environmental Hazards; Trade and Globalisation; Tourism; and Health.

What you will be tested on

For marks to be given, points must relate to the question asked.

There are six types of question used in this paper:

A. Describe
B. Explain
C. Give reasons
D. Match
E. Give map evidence
F. Give advantages and/or disadvantages

Questions which ask candidates to **describe**:

You must make a number of relevant, factual points. These should be key points taken from a given source, for example a map, diagram or table.

Questions which ask candidates to **explain** or **give reasons**:

You should make a number of points giving clear reasons for a given situation. The command word "explain" will be used when you are asked to demonstrate knowledge and understanding. Sometimes the command words "give reasons" may be used as an alternative to "explain".

Questions which ask candidates to **match**:

You are asked to match two sets of variables, for example to match features to a correct grid reference.

Questions which ask candidates to **give map evidence**:

You should look for evidence on the map and make clear statements to support your answer.

Questions which ask candidates to **give advantages and/or disadvantages**:

You should select relevant advantages or disadvantages of a proposed development, for example the location of a new shopping centre, and demonstrate your understanding of the significance of the proposal.

Some tips for revising

- To be best prepared for the examination, organise your notes into sections. Try to work out a schedule for studying with a programme which includes the sections of the syllabus you intend to study.
- Organise your notes into checklists and revision cards.
- Try to avoid leaving your studying to a day or two before the exam. Also try to avoid cramming your studies into the night before the examination, and especially avoid staying up late to study.
- One useful technique when revising is to use summary note cards on individual topics.

- Make use of specimen and model paper questions to test your knowledge or enquiry skills. Go over your answers and give yourself a mark for every correct point you make when comparing your answer with your notes.
- If you work with a classmate, try to mark each other's practice answers.
- Practise your diagram-drawing skills and your writing skills. Ensure that your answers are clearly worded. Try to develop the points that you make in your answers.

Some tips for the exam

- Do not write lists, even if you are running out of time. You will lose marks. If the question asks for an opinion based on a choice, for example on the suitability of a particular site or area for a development, do not be afraid to refer to negative points such as why the alternatives are not as good. You will get credit for this.
- Make sure you have a copy of the examination timetable and have planned a schedule for studying.
- Arrive at the examination in plenty of time with the appropriate equipment – pen, pencil, rubber and ruler.
- Carefully read the instructions on the paper and at the beginning of each part of the question.
- Answer all of the compulsory questions in each paper you sit.
- Use the number of marks as a guide to the length of your answer.
- Try to include examples in your answer wherever possible. If asked for diagrams, draw clear, labelled diagrams.
- Read the question instructions very carefully. If the question asks you to "describe", make sure that this is what you do.
- If you are asked to "explain", you must use phrases such as "due to", "this happens because" and "this is a result of". If you describe rather than explain, you will lose most of the marks for that question.
- If you finish early, do not leave the exam. Use the remaining time to check your answers and go over any questions which you have partially answered, especially Ordnance Survey map questions.
- Practise drawing diagrams which may be included in your answers, for example corries or pyramidal peaks.
- Make sure that you have read the instructions on the question carefully and that you have avoided needless errors. For example, answering the wrong sections or failing to explain when asked to, or perhaps omitting to refer to a named area or case study.

- One technique which you might find helpful, especially when answering long questions worth 10 or more marks, is to "brainstorm" possible points for your answer. You can write these down in a list at the start of your answer. As you go through your answer, you can double-check with your list to ensure that you have put as much into your answer as you can. This stops you from coming out of the exam and being annoyed that you forgot to mention an important point.

Common errors

Markers of the external examination often remark on errors which occur frequently in candidates' answers. These include the following:

Lack of sufficient detail

- Many candidates fail to provide sufficient detail in answers, often by omitting reference to specific examples, or not elaborating or developing points made in their answer. As noted above, a good guide to the amount of detail required is the number of marks given for the question. If, for example, the total marks offered is 6, then you should make at least six valid points.

Listing

- If you write a simple list of points rather than fuller statements in your answer, you will automatically lose marks. For example, in a 4 mark question, you will obtain only 1 mark for a list.
- The same rule applies to a simple list of bullet points. However, if you couple bullet points with some detailed explanation, you could achieve full marks.

Irrelevant answers

- You must read the question instructions carefully so as to avoid giving answers which are irrelevant to the question. For example, if you are asked to "explain" and you simply "describe", you will lose marks. If you are asked for a named example and you do not provide one, you will forfeit marks.

Repetition

- You should be careful not to repeat points already made in your answer. These will not gain any further marks. You may feel that you have written a long answer, but it may contain the same basic information repeated again and again. Unfortunately, these repeated statements will be ignored by the marker.

Good luck!

Remember that the rewards for passing National 5 Geography are well worth it! Your pass will help you to get the future you want for yourself. In the exam, be confident in your own ability. If you're not sure how to answer a question, trust your instincts and just give it a go anyway. Keep calm and don't panic! GOOD LUCK!

Model Paper 1

Whilst this Model Paper has been specially commissioned by Hodder Gibson for use as practice for the National 5 exams, the key reference documents remain the SQA Specimen Paper 2013 and the SQA Past Papers 2014 and 2015.

HODDER GIBSON
LEARN MORE

N5

National
Qualifications
MODEL PAPER 1

Geography

Duration — 1 hour and 30 minutes

Total marks — 60

SECTION 1 — PHYSICAL ENVIRONMENTS — 20 marks

Attempt EITHER question 1 **or** question 2 AND questions 3, 4 and 5

SECTION 2 — HUMAN ENVIRONMENTS — 20 marks

Attempt questions 6, 7 and 8

SECTION 3 — GLOBAL ISSUES — 20 marks

Attempt any TWO of the following

Question   9 — Climate Change

Question 10 — Impact of Human Activity on the Natural Environment

Question 11 — Environmental Hazards

Question 12 — Trade and Globalisation

Question 13 — Tourism

Question 14 — Health

You should read the questions carefully.

Credit will always be given for appropriately labelled sketch maps and diagrams.

Use **blue** or **black** ink.

HODDER
GIBSON
LEARN MORE

OS MAP ITEM A

Extract No 1655/36

1:50 000 Scale
Landranger Series

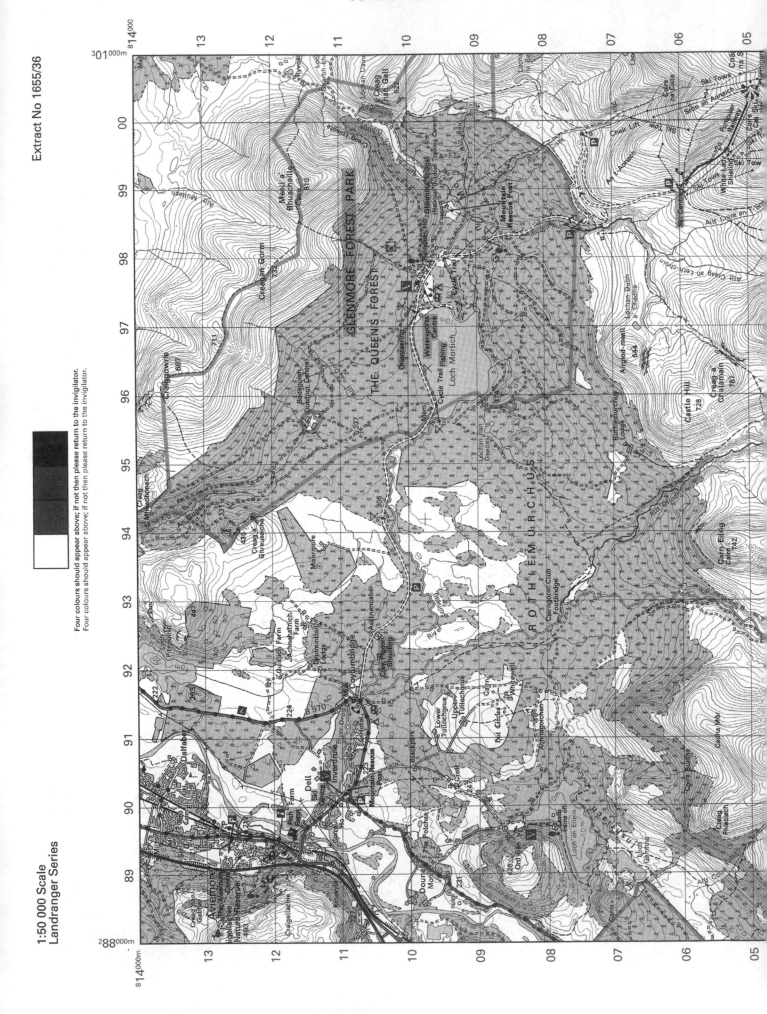

Scale 1: 50 000

2 centimetres to 1 kilometre (one grid square)

1 kilometre = 0·6214 mile

1 mile = 1·6093 kilometres

OS MAP ITEM B

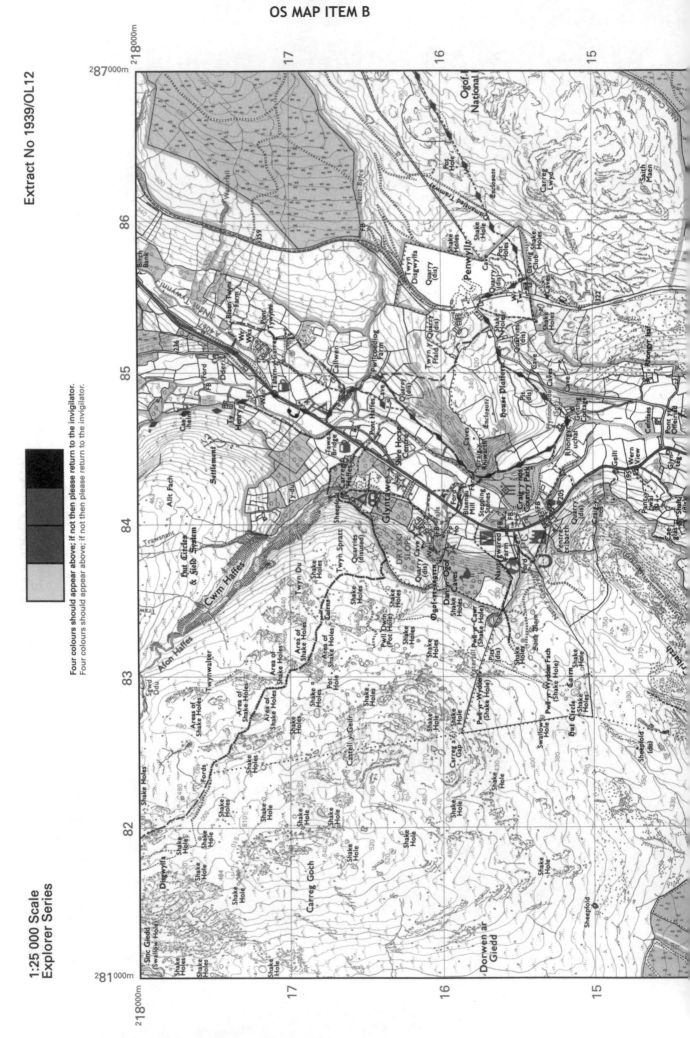

Extract No 1939/OL12

1:25 000 Scale
Explorer Series

Four colours should appear above; if not then please return to the invigilator.
Four colours should appear above; if not then please return to the invigilator.

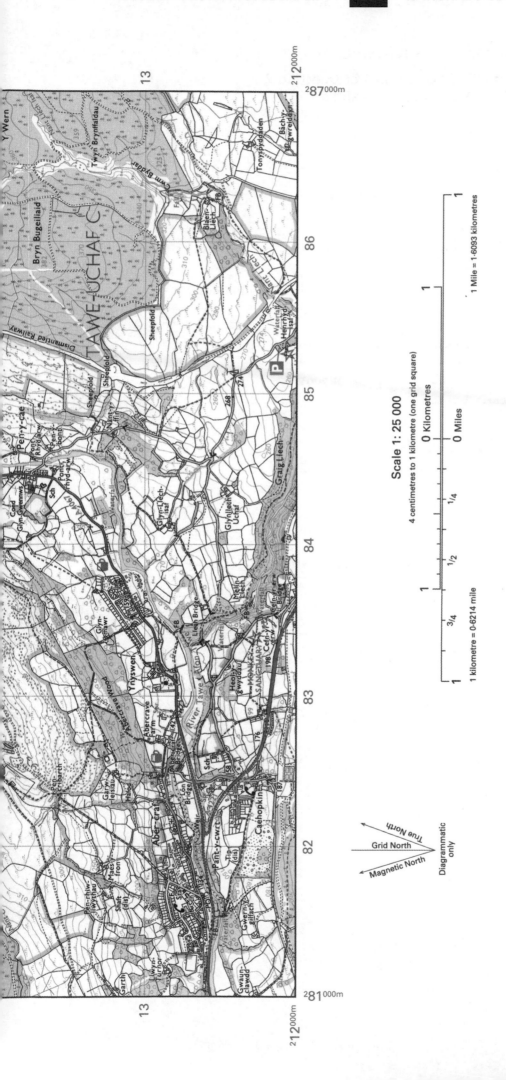

Scale 1: 25 000

4 centimetres to 1 kilometre (one grid square)

1 kilometre = 0·6214 mile

1 Mile = 1·6093 kilometres

Diagrammatic only

Grid North
True North
Magnetic North

MARKS | DO NOT WRITE IN THIS MARGIN

SECTION 1 — PHYSICAL ENVIRONMENTS — 20 marks

Attempt EITHER Question 1 or Question 2
AND Questions 3, 4 and 5

Question 1 — Glaciated Uplands

(a) Study OS map **Item A** of the Aviemore area.

Match the features of glaciated uplands shown below with the correct grid reference.

Features of glaciated uplands		
U-shaped valley	corrie	pyramidal peak

Choose from grid references			
954976	979976	957997	915980

3

(b) **Explain** the formation of a U-shaped valley.

You may use a diagram or diagrams in your answer. **4**

Total marks **7**

NOW ATTEMPT QUESTIONS 3, 4 AND 5

MARKS | DO NOT WRITE IN THIS MARGIN

DO NOT ATTEMPT THIS QUESTION IF YOU HAVE ALREADY ANSWERED QUESTION 1

Question 2 — Upland Limestone Areas

(a) Study OS map **Item B** of the Aber-craf area.

Match the following surface limestone features with the correct grid reference.

Features of upland limestone areas		
pot hole	limestone pavement	intermittent drainage

Choose from grid references			
833164	817155	819163	814175

3

(b) **Explain** the formation of limestone pavement.

You may use a diagram or diagrams in your answer.

4

Total marks   7

NOW ATTEMPT QUESTIONS 3, 4 AND 5

MARKS | DO NOT WRITE IN THIS MARGIN

NOW ATTEMPT QUESTIONS 3, 4 AND 5

Question 3

Study OS map **Item A** of the Aviemore area.

Referring to map evidence, **explain** the ways in which the physical landscape has affected land use in the map extract area.

5

Question 4

Diagram Q4 — Synoptic Chart for 30 May 2013

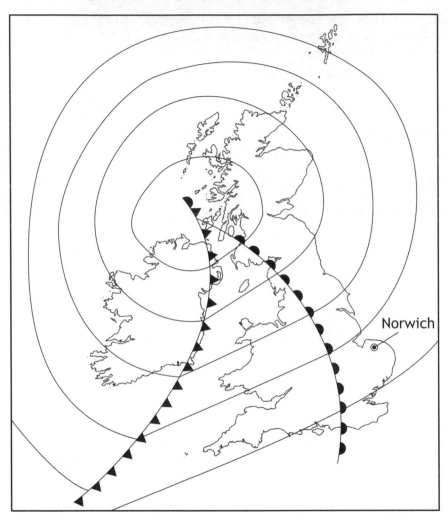

Study Diagram Q4 above.

Explain the changes that will take place in the weather at Norwich over the next twenty-four hours.

4

MARKS | DO NOT WRITE IN THIS MARGIN

Question 5

Diagram Q5A — Landscape Types

Diagram Q5B — Land Uses

Glaciated Uplands

Upland Limestone

Farming

Forestry

Industry

Recreation and Tourism

Water Storage and Supply

Renewable Energy

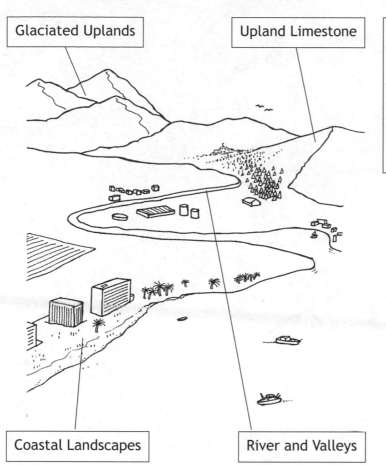

Coastal Landscapes

River and Valleys

Look at Diagrams Q5A and Q5B.

Choose one landscape type you have studied from Diagram Q5A.

Select at least **two** land uses from Diagram Q5B and **explain** why these land uses are suitable for your chosen landscape type.

4

OS MAP ITEM C

Extract No 1879/EXP308

1:25 000 Scale
Explorer Series

Four colours should appear above; if not then please return to the invigilator.
Four colours should appear above; if not then please return to the invigilator.

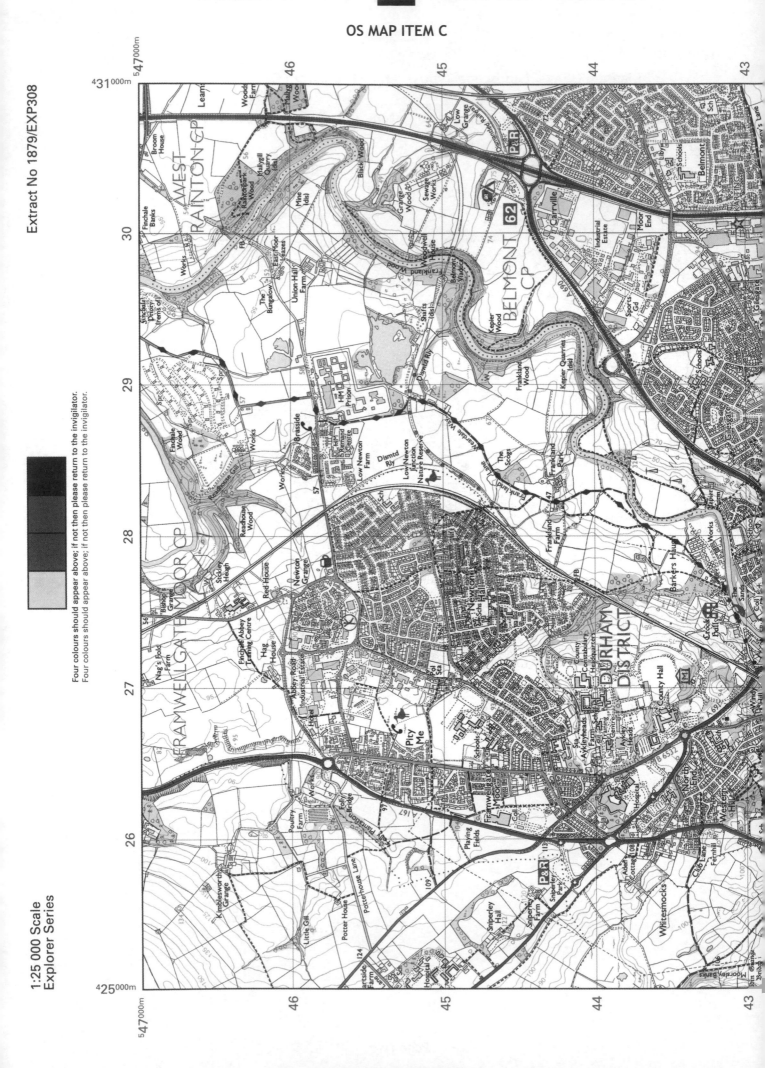

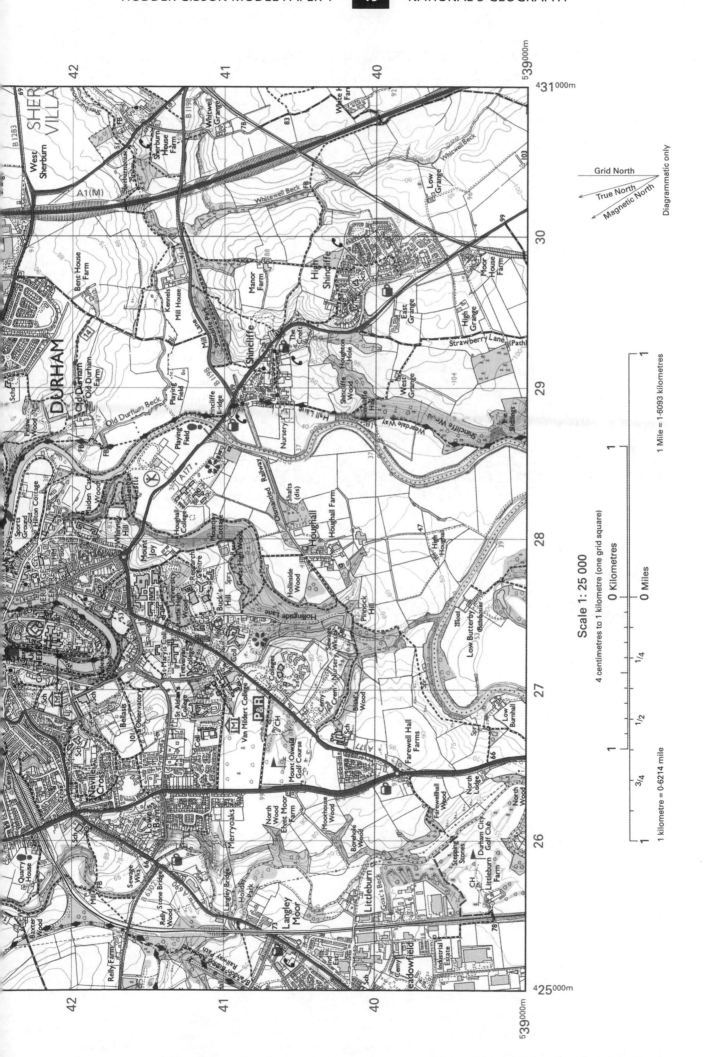

Scale 1: 25 000

4 centimetres to 1 kilometre (one grid square)

1 kilometre = 0·6214 mile

1 Mile = 1·6093 kilometres

Grid North

True North

Magnetic North

Diagrammatic only

MARKS | DO NOT WRITE IN THIS MARGIN

SECTION 2 — HUMAN ENVIRONMENTS — 20 marks
Attempt Questions 6, 7 and 8

Question 6

Study OS map **Item C** of the Durham area.

    (a) **Using Item C, give map evidence** to show that part of Durham's CBD is located in grid square 2742.   **3**

    (b) **Describe**, **in detail**, differences between the urban environments in grid squares 2642 and 2745.   **5**

Question 7

Diagram Q7 — Complexo do Alemão Shanty Town, Rio de Janeiro

Population 100 000

One of the biggest slums in Rio

Housing made from cardboard, mud and tin sheets

Open sewers and toilets

No electricity

Few roads

Look at Diagram Q7.

For Rio de Janeiro, or a named developing world city you have studied, **describe** methods used by city authorities to improve living conditions in shanty towns.   **6**

Question 8

Diagram Q8 — Modern Developments in Farming

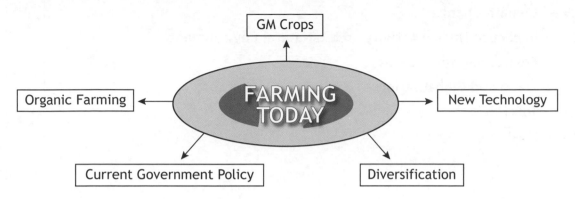

Look at Diagram Q8.

Choose **two** modern developments from Diagram Q8.

Explain how your chosen developments affect farmers in the developed world.   6

NOW GO TO SECTION 3

MARKS | DO NOT WRITE IN THIS MARGIN

SECTION 3 — GLOBAL ISSUES — 20 marks

Attempt any TWO questions

Question 9 — Climate Change

Question 10 — Impact of Human Activity on the Natural Environment

Question 11 — Environmental Hazards

Question 12 — Trade and Globalisation

Question 13 — Tourism

Question 14 — Health

MARKS | DO NOT WRITE IN THIS MARGIN

Question 9 — Climate Change

Diagram Q9 — Ice Melt 1979–2012 (months of June and July only)

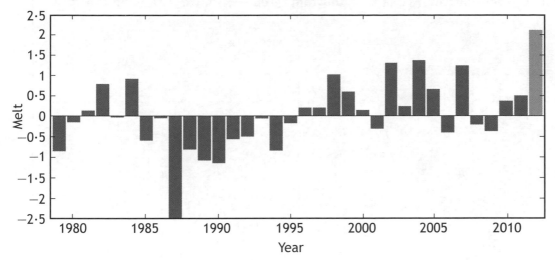

Study Diagram Q9.

(a) **Describe**, in detail, changes in ice melt between 1979 and 2012.   4

(b) Explain the effects of climate change on people and the environment.   6

Total marks 10

Question 10 — Impact of Human Activity on the Natural Environment

Diagram Q10A — Barrow, Alaska   Diagram Q10B — Eismitte, Greenland

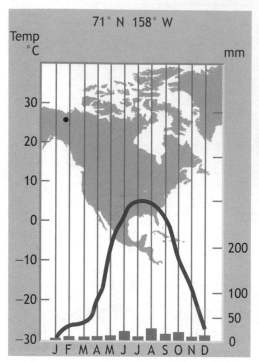 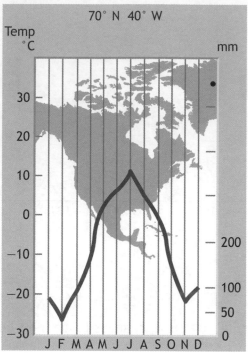

Study Diagrams Q10A and Q10B.

(a) **Describe, in detail,** differences between the climates of Barrow and Eismitte.

4

Diagram Q10C — Land Use in the Rainforest

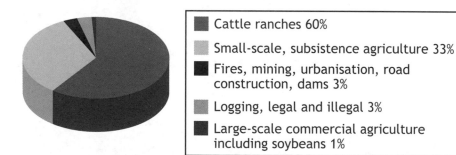

- Cattle ranches 60%
- Small-scale, subsistence agriculture 33%
- Fires, mining, urbanisation, road construction, dams 3%
- Logging, legal and illegal 3%
- Large-scale commercial agriculture including soybeans 1%

(b) Look at Diagram Q10C.

Explain how two of the land uses shown in Diagram Q10C can lead to degradation of the rainforest.

6

Total marks 10

Question 11 — Environmental Hazards

Diagram Q11A — The World's Most Active Volcanoes

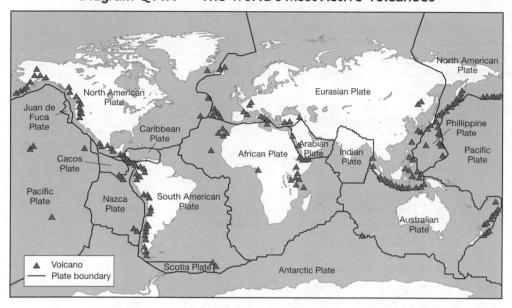

Study Diagram Q11A.

(a) **Describe, in detail,** the distribution of the world's most active volcanoes.   **4**

Diagram Q11B — Environmental Hazards

Look at Diagram Q11B.

(b) For a named earthquake, volcano or tropical storm you have studied, **explain** the impact it had on the people and the environment.   **6**

Total marks   **10**

MARKS

Question 12 — Trade and Globalisation

Diagram Q12A — Percentage Share of World Goods Production

1990	Developed countries	70%
	China	4%
	India	4%
2000	Developed countries	68%
	China	7%
	India	4%
2010	Developed countries	50%
	China	14%
	India	6%
2016 (projected)	Developed countries	45%
	China	18%
	India	7%

Study Diagram Q12A.

(a) **Describe, in detail,** the trends in percentage share of world goods production.

4

Diagram Q12B — Newspaper Headline

"The inequality in trade between rich and poor nations is now wider than it has ever been before"

Look at Diagram Q12B.

(b) **Explain** the causes of inequalities in trade between developed and developing countries.

6

Total marks 10

MARKS | DO NOT WRITE IN THIS MARGIN

Question 13 — Tourism

Diagram Q13 — Global Tourist Arrivals 1990–2011

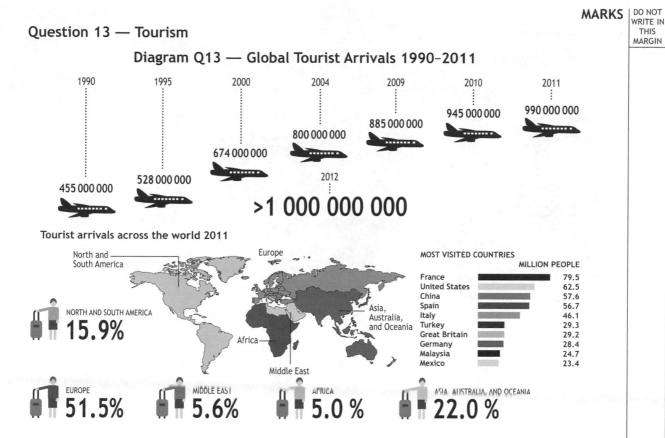

| 1990 | 1995 | 2000 | 2004 | 2009 | 2010 | 2011 |

945 000 000   990 000 000

885 000 000

800 000 000

674 000 000

2012

>1 000 000 000

455 000 000   528 000 000

Tourist arrivals across the world 2011

North and South America   Europe

NORTH AND SOUTH AMERICA
15.9%

Asia, Australia, and Oceania

Africa

Middle East

MOST VISITED COUNTRIES

	MILLION PEOPLE
France	79.5
United States	62.5
China	57.6
Spain	56.7
Italy	46.1
Turkey	29.3
Great Britain	29.2
Germany	28.4
Malaysia	24.7
Mexico	23.4

EUROPE
51.5%

MIDDLE EAST
5.6%

AFRICA
5.0 %

ASIA, AUSTRALIA, AND OCEANIA
22.0 %

Study Diagram Q13.

(a) **Describe** the trends in global tourist arrivals between 1990 and 2011.   4

(b) **Explain** the advantages and disadvantages of mass tourism.   6

Total marks 10

Question 14 — Health

Diagram Q14A — Adults Infected with HIV/AIDS

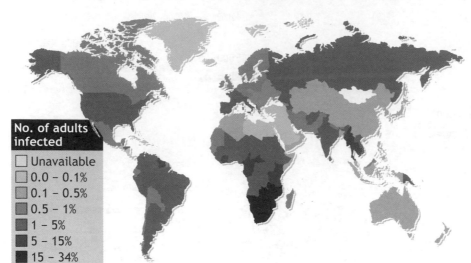

No. of adults infected
☐ Unavailable
☐ 0.0 – 0.1%
☐ 0.1 – 0.5%
☐ 0.5 – 1%
☐ 1 – 5%
☐ 5 – 15%
■ 15 – 34%

Study Diagram Q14A.

(a) **Describe, in detail,** the distribution of HIV/AIDS.   4

Diagram Q14B — Selected Health Facts for Scotland (2011)

More than 8000 people die from heart disease each year

It is estimated that more than 1 in 3 people in Scotland will develop some form of cancer during their lifetime

In Scotland, 368 000 people are currently receiving treatment for asthma.

Study Diagram Q14B.

(b) (i) **Explain** the main causes of heart disease, **or** cancer, **or** asthma.   3

(ii) For your chosen disease in part (i), **describe** some methods used to help prevent **or** reduce the disease.   3

Total marks 10

[END OF MODEL PAPER]

Model Paper 2

Whilst this Model Paper has been specially commissioned by Hodder Gibson for use as practice for the National 5 exams, the key reference documents remain the SQA Specimen Paper 2013 and the SQA Past Papers 2014 and 2015.

HODDER GIBSON

LEARN MORE

N5

National Qualifications MODEL PAPER 2

Geography

Duration — 1 hour and 30 minutes

Total marks — 60

SECTION 1 — PHYSICAL ENVIRONMENTS — 20 marks

Attempt EITHER question 1 **or** question 2 AND questions 3, 4 and 5

SECTION 2 — HUMAN ENVIRONMENTS — 20 marks

Attempt questions 6, 7 and 8

SECTION 3 — GLOBAL ISSUES — 20 marks

Attempt any TWO of the following

Question   9 — Climate Change

Question 10 — Impact of Human Activity on the Natural Environment

Question 11 — Environmental Hazards

Question 12 — Trade and Globalisation

Question 13 — Tourism

Question 14 — Health

You should read the questions carefully.

Credit will always be given for appropriately labelled sketch maps and diagrams.

Use **blue** or **black** ink.

HODDER
GIBSON
LEARN MORE

OS MAP ITEM A

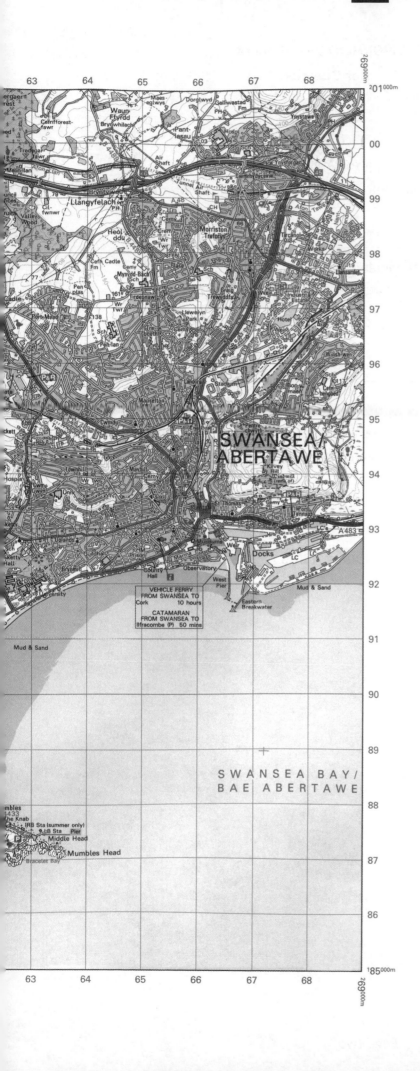

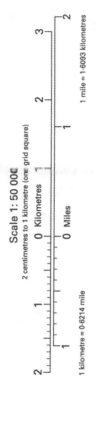

Scale 1: 50 000

2 centimetres to 1 kilometre (one grid square)

1 mile = 1·6093 kilometres

1 kilometre = 0·6214 mile

True North

Grid North

Magnetic North

Diagrammatic only

MARKS | DO NOT WRITE IN THIS MARGIN

SECTION 1 — PHYSICAL ENVIRONMENTS — 20 marks

**Attempt EITHER Question 1 or Question 2
AND Questions 3, 4 and 5**

Question 1 — Coastal Landscapes

(a) Study OS map **Item A** of the Swansea area.

Match the coastal features shown below with the correct grid reference.

Features of coastal landscapes		
bay	headland	stack

Choose from grid references			
636871	570863	592974	555869

3

(b) **Explain** the formation of caves, arches and stacks.

You may use a diagram or diagrams in your answer.   **4**

Total marks   7

NOW ATTEMPT QUESTIONS 3, 4 AND 5

MARKS | DO NOT WRITE IN THIS MARGIN

DO NOT ATTEMPT THIS QUESTION IF YOU HAVE ALREADY ANSWERED QUESTION 1

Question 2 — Rivers and Valleys

(a) Study OS map **Item A** of the Swansea area.

Match the river and valley features shown below with the correct grid reference.

Features of rivers and valleys		
confluence	meander	river flowing south

Choose from grid references			
670967	672973	584973	614965

3

(b) **Explain** the formation of a waterfall.

You may use a diagram or diagrams in your answer.

4

Total marks 7

NOW ATTEMPT QUESTIONS 3, 4 AND 5

NOW ATTEMPT QUESTIONS 3, 4 AND 5

Question 3

Study OS map **Item A** of the Swansea area.

The area south of Swansea is a popular destination for tourists.

Using map evidence, explain the attractions of the physical landscape for tourists.   **5**

Question 4

Diagram Q4 — An Anticyclone over the UK

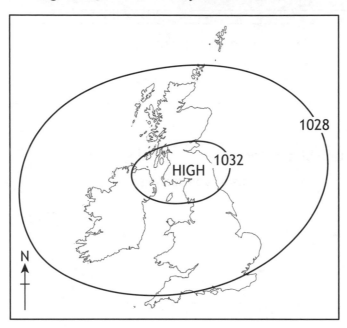

Look at Diagram Q4.

Describe similarities and differences in weather conditions caused by anticyclones in summer and winter.   **4**

MARKS | DO NOT WRITE IN THIS MARGIN

Question 5

Diagram Q5 — Land-Use Types and Land Uses

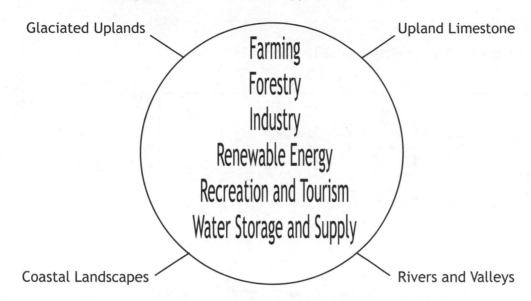

Glaciated Uplands

Upland Limestone

Farming
Forestry
Industry
Renewable Energy
Recreation and Tourism
Water Storage and Supply

Coastal Landscapes

Rivers and Valleys

Look at Diagram Q5.

Choose one landscape type which you have studied.

For your chosen landscape, **describe** ways in which two of the land uses shown are in conflict with each other.

4

MARKS | DO NOT WRITE IN THIS MARGIN

SECTION 2 — HUMAN ENVIRONMENTS — 20 marks

Attempt Questions 6, 7 and 8

Question 6

Diagram Q6 — Selected Land-Use Zones in Swansea

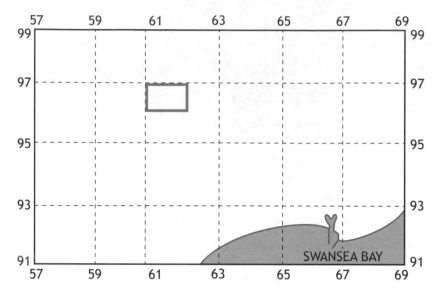

Study OS map **Item A** of the Swansea area and Diagram Q6.

It is proposed to build a shopping complex in square 6196. **Using map evidence, describe** the advantages and disadvantages of this site.   **5**

Question 7

Diagram Q7 — World Population Distribution

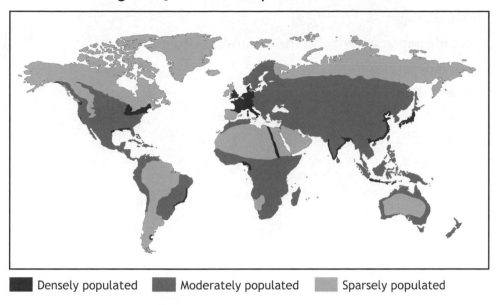

■ Densely populated   ■ Moderately populated   ■ Sparsely populated

Study Diagram Q7.

(a) **Describe**, in detail, the distribution of the world's population.   3

(b) **Explain** factors which influence the distribution of the world's population.

You should refer to **both** human and physical factors in your answer.   6

Total marks   9

Question 8

Diagram Q8 — Indicators of Development in the USA and Chad

Indicators	USA — Developed Country	Chad — Developing Country
Birth rate	13	40
Exports	steel, motor vehicles, telecommunications	oil, cattle, cotton
Life expectancy at birth	79	50
Adult literacy rate	99%	35%
GDP per capita (US $)	$49 000	$2 000

Study Diagram Q8.

Choose **two** indicators from Diagram Q8.

Explain the reasons for the different levels of development in the USA and Chad.   6

NOW GO TO SECTION 3

SECTION 3 — GLOBAL ISSUES — 20 marks

Attempt any TWO questions

Question 9 — Climate Change

Question 10 — Impact of Human Activity on the Natural Environment

Question 11 — Environmental Hazards

Question 12 — Trade and Globalisation

Question 13 — Tourism

Question 14 — Health

MARKS | DO NOT WRITE IN THIS MARGIN

Question 9 — Climate Change

Diagram Q9 — Average Global Temperatures 1880–2020

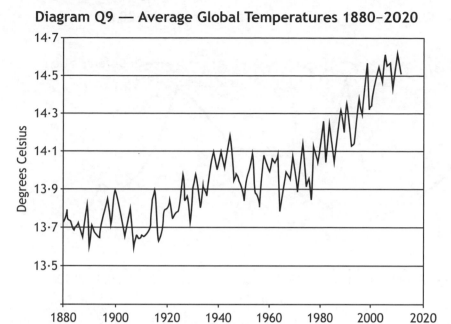

Study Diagram Q9.

(a) **Describe, in detail,** the changes in average global temperatures between 1880 and 2020 (predicted). **4**

(b) **Explain** the physical and human causes of climate change. **6**

Total marks 10

MARKS | DO NOT WRITE IN THIS MARGIN

Question 10 — Impact of Human Activity on the Natural Environment

Diagram Q10 — Amazon Deforestation per Year, 1988–2010

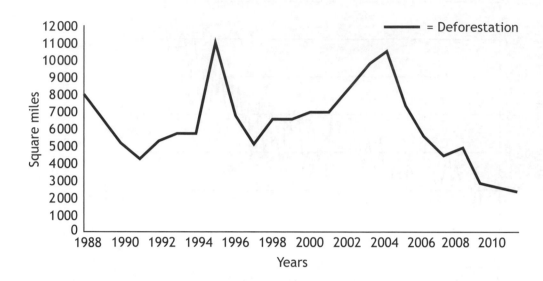

(a) Study Diagram Q10.

Describe, in detail, the trends in deforestation between 1988 and 2010.   **4**

(b) **Explain** ways in which the destruction of the rainforest can be reduced.   **6**

Total marks 10

Question 11 — Environmental Hazards

Diagram Q11 — Distribution of Tropical Storms

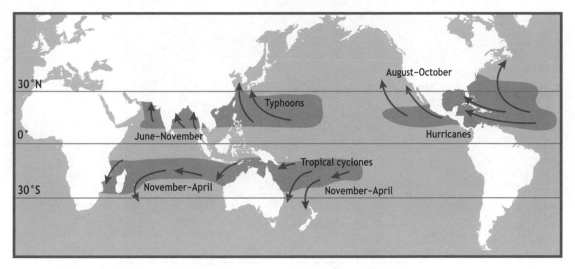

Study Diagram Q11.

(a) **Describe, in detail,** the distribution of tropical storms.   4

(b) Select **one** type of natural disaster from earthquakes, volcanoes or tropical storms.

Explain methods used to predict and to plan for their occurrence.   6

Total marks 10

Question 12 — Trade and Globalisation

Diagram Q12A — Share of World Exports

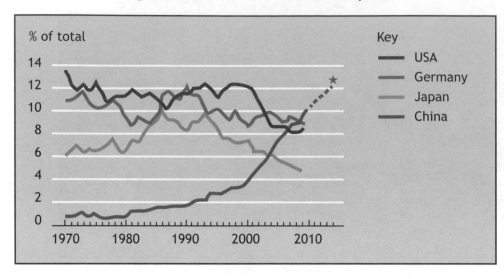

(a) Study Diagram Q12A.

Describe, in detail, the changes in the share of world exports between 1970 and 2010.

4

Diagram Q12B — Exports of a Developing Country

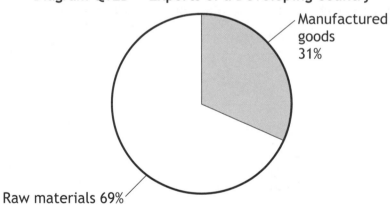

Manufactured goods 31%

Raw materials 69%

(b) Look at Diagram Q12B.

Many developing countries rely on one product to export. **Explain** the effects that changing demand for the product will have on the people and the environment.

6

Total marks 10

MARKS

Question 13 — Tourism

Diagram Q13 — Facts on Selected National Parks in the UK

National Park name	Year of designation	Population	Visitors per year (million)	Visitor days per year (million)	Visitor spend per year (million)
Brecon Beacons	1957	32 000	4·15	5	£197
Cairngorms	2003	17 000	1·5	3·1	£185
Lake District	1951	42 200	15·8	23·1	£952
Loch Lomond and the Trossachs	2002	15 600	4	7	£190
Pembrokeshire coast	1952	22 600	4·2	13	£498
Snowdonia	1951	25 482	4·27	10·4	£396

Study Diagram Q13.

(a) **Describe, in detail,** the differences between the selected National Parks.   4

(b) For a named area you have studied, **explain** methods used to manage problems caused by mass tourism.   6

Total marks 10

MARKS

Question 14 — Health

Diagram Q14 — Cholera Outbreaks 2010–2011

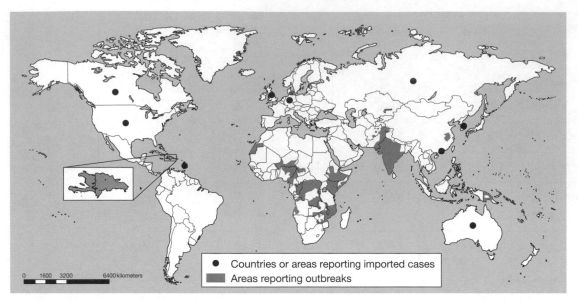

Study Diagram Q14.

(a) **Describe, in detail,** the distribution of cholera outbreaks in 2010 and 2011.

4

(b) For cholera, malaria, kwashiorkor **or** pneumonia, **explain** the main causes of your chosen disease and **describe** some methods used to help prevent or reduce the disease.

6

Total marks 10

[END OF MODEL PAPER]

Model Paper 3

Whilst this Model Paper has been specially commissioned by Hodder Gibson for use as practice for the National 5 exams, the key reference documents remain the SQA Specimen Paper 2013 and the SQA Past Papers 2014 and 2015.

HODDER GIBSON
LEARN MORE

N5

National
Qualifications
MODEL PAPER 3

Geography

Duration — 1 hour and 30 minutes
Total marks — 60

SECTION 1 — PHYSICAL ENVIRONMENTS — 20 marks

Attempt EITHER question 1 **or** question 2 AND questions 3, 4 and 5

SECTION 2 — HUMAN ENVIRONMENTS — 20 marks

Attempt questions 6, 7 and 8

SECTION 3 — GLOBAL ISSUES — 20 marks

Attempt any TWO of the following

Question   9 — Climate Change

Question 10 — Impact of Human Activity on the Natural Environment

Question 11 — Environmental Hazards

Question 12 — Trade and Globalisation

Question 13 — Tourism

Question 14 — Health

You should read the questions carefully.

Credit will always be given for appropriately labelled sketch maps and diagrams.

Use **blue** or **black** ink.

OS MAP ITEM A

Extract No 1880/20

1:50 000 Scale
Landranger Series

Four colours should appear above; if not then please return to the invigilator.
Four colours should appear above; if not then please return to the invigilator.

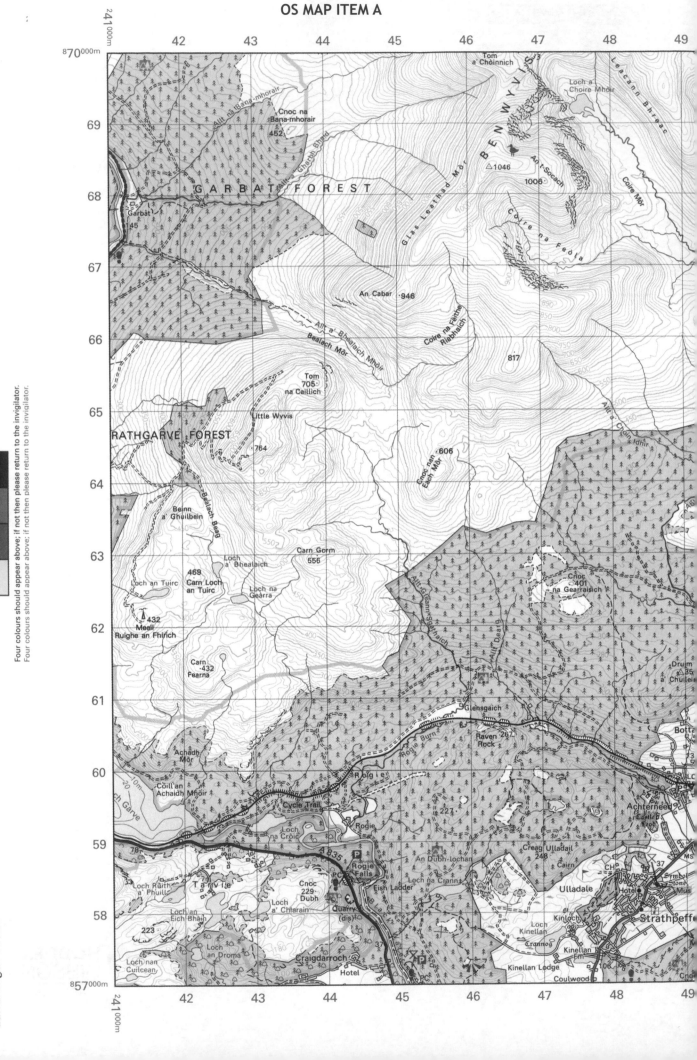

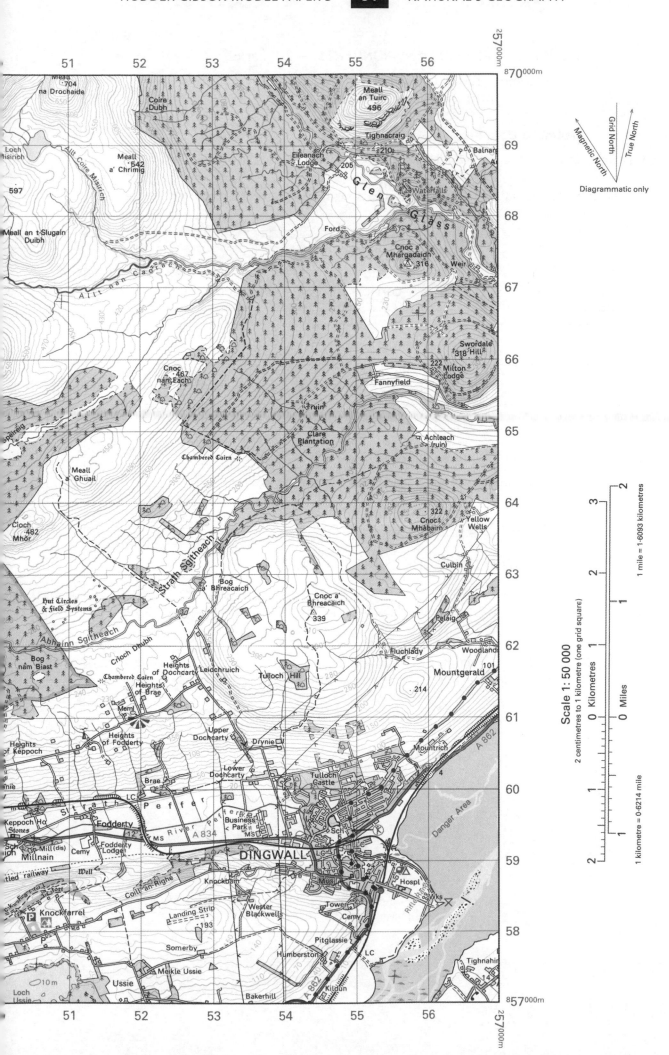

Grid North

Magnetic North

True North

Diagrammatic only

Scale 1:50 000

2 centimetres to 1 kilometre (one grid square)

1 kilometre = 0·6214 mile

1 mile = 1·6093 kilometres

MARKS

SECTION 1 — PHYSICAL ENVIRONMENTS — 20 marks

**Attempt EITHER Question 1 or Question 2
AND Questions 3, 4 and 5**

Question 1 — Glaciated Uplands

Study OS map **Item A** of the Dingwall area.

(a) **Match** the glaciated uplands features shown below with the correct grid reference.

Features of glaciated uplands		
arête	corrie	U-shaped valley

Choose from grid references			
467677	472693	435663	525594

3

(b) **Explain** the formation of a corrie.

You may use a diagram or diagrams in your answer.   4

Total marks   7

NOW ATTEMPT QUESTIONS 3, 4 AND 5

MARKS | DO NOT WRITE IN THIS MARGIN

DO NOT ATTEMPT THIS QUESTION IF YOU HAVE ALREADY ANSWERED QUESTION 1

Question 2 — Rivers and Valleys

Study OS map **Item A** of the Dingwall area.

(a) **Match** the following river and valley features to the correct grid references.

Features of rivers and valleys		
V-shaped valley	meander	tributary

Choose from grid references			
528595	442591	447585	473657

3

(b) **Explain** the formation of a V-shaped valley.

You may use a diagram or diagrams in your answer.

4

Total marks   7

NOW ATTEMPT QUESTIONS 3, 4 AND 5

MARKS | DO NOT WRITE IN THIS MARGIN

NOW ATTEMPT QUESTIONS 3, 4 AND 5

Question 3

Diagram Q3A — Cross-section from GR 466658 to GR 510580

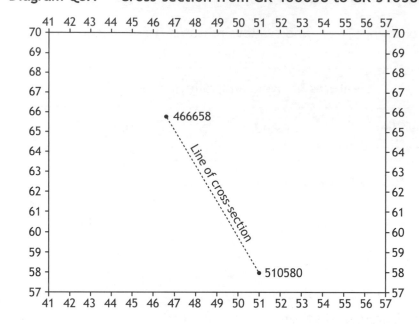

Diagram Q3B — Cross-section from GR 466658 to GR 510580 Showing Land Use

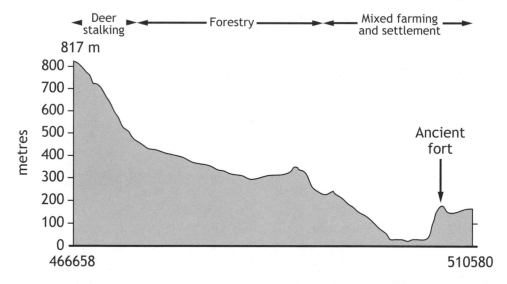

Look at the map extract and find the cross-section shown on Diagrams Q3A and Q3B.

Study both the map extract and the cross-section.

Give reasons for the differences in land use along the section.

5

Question 4

Diagram Q4 — Synoptic Chart, 26 February 2013

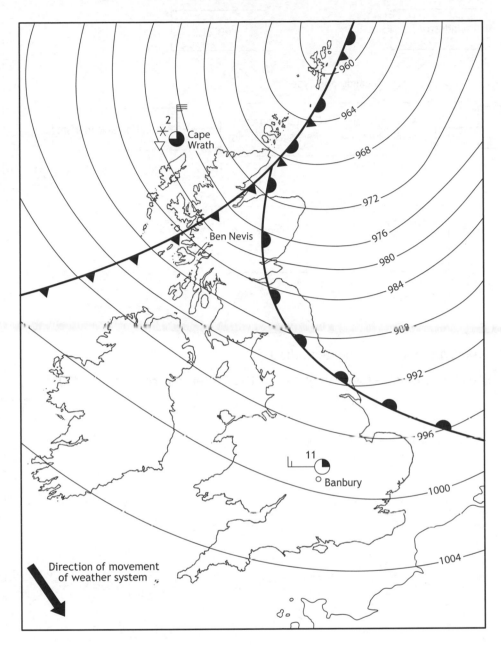

Study Diagram Q4.

Use the information in Diagram Q4 to **give reasons** for the differences in the weather conditions between Banbury and Cape Wrath.

4

Question 5

Diagram Q5 — Landscape Types and Selected Land Uses

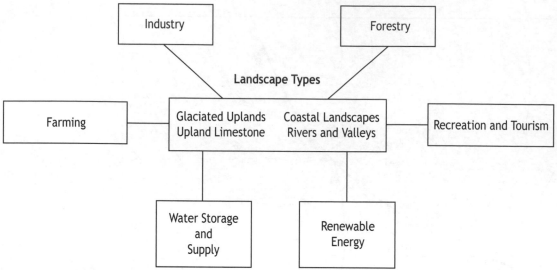

Look at Diagram Q5.

Choose one landscape type which you have studied.

For your chosen landscape, **give reasons** why one of the land uses from Diagram Q5 is found there.

4

MARKS | DO NOT WRITE IN THIS MARGIN

SECTION 2 — HUMAN ENVIRONMENTS — 20 marks
Attempt Questions 6, 7 and 8

Question 6

Study OS map **Item A** of the Dingwall area.

There is a plan to build a new housing estate in grid square 5558.

Using map evidence, give the advantages and disadvantages of this proposal.   **5**

Question 7

Diagram Q7 — GDP per Capita and Number of Births per Woman in Selected Countries

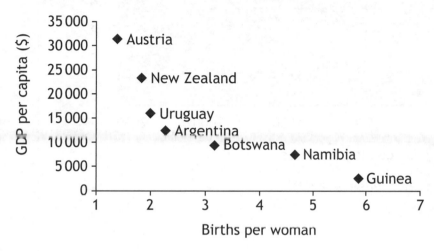

Study Diagram Q7.

    (a) **Describe** the link between GDP per capita and the number of births per woman as shown in the graph.   **3**

"GDP per capita is a measure of wealth".

    (b) **Explain** why there is a link between GDP per capita and the number of births per woman as shown in the graph.   **6**

Total marks   9

MARKS | DO NOT WRITE IN THIS MARGIN

Question 8

Diagram Q8 — World Population Growth (projected)

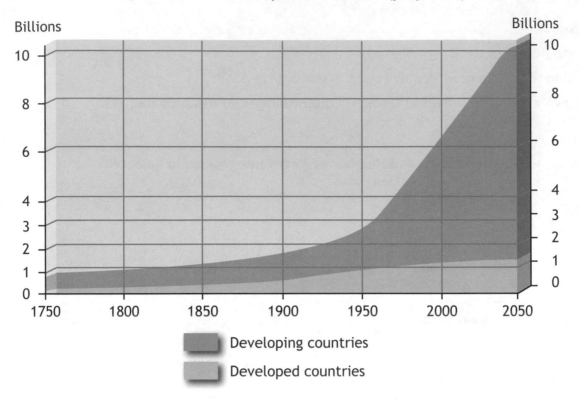

Billions (left axis) / Billions (right axis)

- Developing countries
- Developed countries

Look at Diagram Q8.

What are the implications of the population growth shown on Diagram Q8 for developing and developed countries?   **6**

NOW GO TO SECTION 3

MARKS | DO NOT WRITE IN THIS MARGIN

SECTION 3 — GLOBAL ISSUES — 20 marks

Attempt any TWO questions

Question 9 — Climate Change

Question 10 — Impact of Human Activity on the Natural Environment

Question 11 — Environmental Hazards

Question 12 — Trade and Globalisation

Question 13 — Tourism

Question 14 — Health

MARKS | DO NOT WRITE IN THIS MARGIN

Question 9 — Climate Change

Diagram Q9A — Changes in Carbon Dioxide and Temperatures, 1900–2010

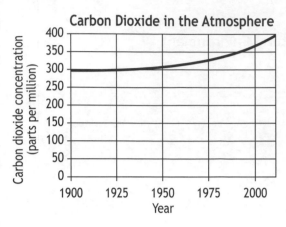
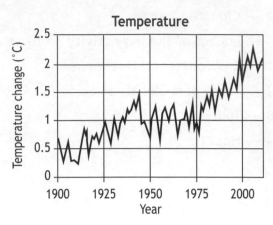

(a) Study Diagram Q9A.

Describe, in detail, changes in carbon-dioxide levels and global temperatures between 1900 and 2010.

4

Diagram Q9B — Some Causes of Climate Change

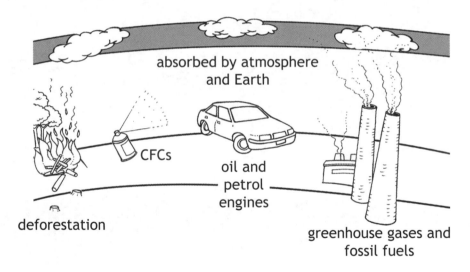

(b) Look at Diagram Q9B.

Explain ways in which people can try to reduce the effects of climate change.

6

Total marks 10

Question 10 — Impact of Human Activity on the Natural Environment

Diagram Q10 — SE Asia: Changes in Forest Cover, 1985–2010

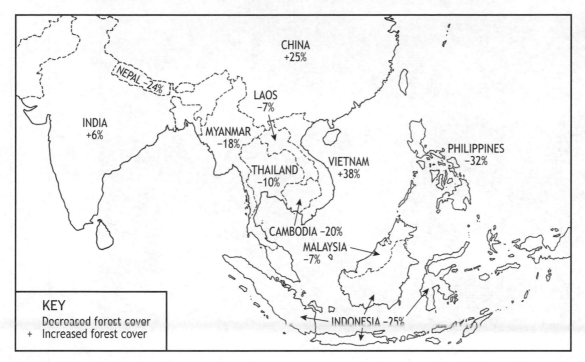

KEY
Decreased forest cover
+ Increased forest cover

Study Diagram Q10.

(a) **Describe, in detail,** changes in forest cover in South-East Asia between 1985–2010.   **4**

(b) **Explain** the advantages and disadvantages of deforestation in equatorial areas.

   In your answer, you should refer to an area you have studied.   **6**

Total marks 10

Question 11 — Environmental Hazards

Diagram Q11 — Distribution of Earthquakes, October 2012

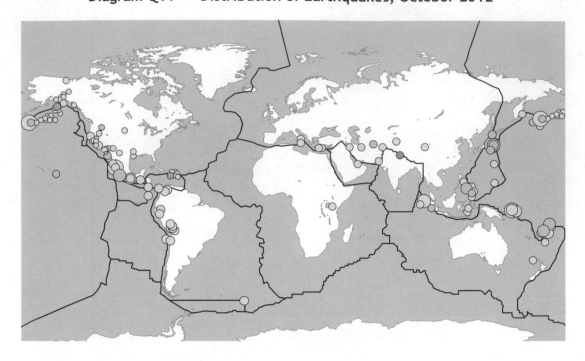

Study Diagram Q11.

(a) **Describe, in detail,** the distribution of earthquakes in October, 2012.   4

(b) For a named environmental hazard you have studied, **explain** the ways in which aid can lessen the impact of environmental hazards.   6

Total marks 10

MARKS | DO NOT WRITE IN THIS MARGIN

Question 12 — Trade and Globalisation

Diagram Q12A — A Selected Country's Imports and Exports

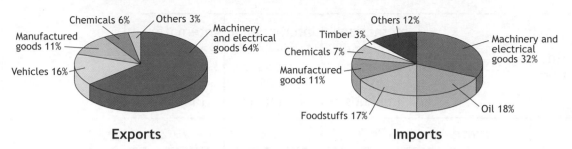

Chemicals 6% Others 3% Machinery and electrical goods 64%
Manufactured goods 11%
Vehicles 16%

Exports

Others 12% Timber 3% Machinery and electrical goods 32%
Chemicals 7%
Manufactured goods 11%
Foodstuffs 17% Oil 18%

Imports

(a) Study Diagram Q12A.

    Describe the patterns of imports and exports shown in Diagram Q12A.   **4**

Diagram Q12B — Main Exports from Selected Countries

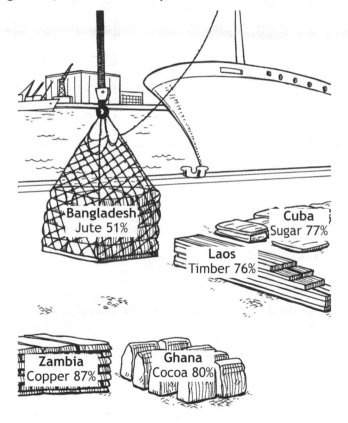

Bangladesh Jute 51%
Cuba Sugar 77%
Laos Timber 76%
Zambia Copper 87%
Ghana Cocoa 80%

(b) Look at Diagram Q12B.

    Explain why the pattern of trade shown in the diagram can cause problems for a developing country.   **6**

                                                        Total marks 10

Question 13 — Tourism

Diagram Q13 — International Tourist Arrivals, 2008–2010

World position	Country	International arrivals of tourists in 2010 (millions)	International arrivals of tourists in 2009 (millions)	International arrivals of tourists in 2008 (millions)
1	France	76·3	76·8	79·2
2	Spain	60·1	55·2	57·9
3	USA	59·7	55·1	57·2
4	China	55·7	50·9	53·0
5	Italy	53·6	43·2	42·7
6	United Kingdom	28·1	28·2	30·1
7	Turkey	27·0	25·5	25·0
8	Germany	26·9	24·2	24·9
9	Malaysia	24·6	23·6	22·1
10	Mexico	22·4	21·5	22·6
Worldwide total		940	882	917

Study Diagram Q13.

(a) **Describe, in detail,** patterns in international tourist arrivals between 2008 and 2010. **4**

(b) **Explain** ways in which eco-tourism helps the people and the environment of a developing country.

In your answer, you should refer to a country you have studied. **6**

Total marks 10

MARKS | DO NOT WRITE IN THIS MARGIN

Question 14 — Health

Diagram Q14 — Malaria Cases in Selected South African Provinces

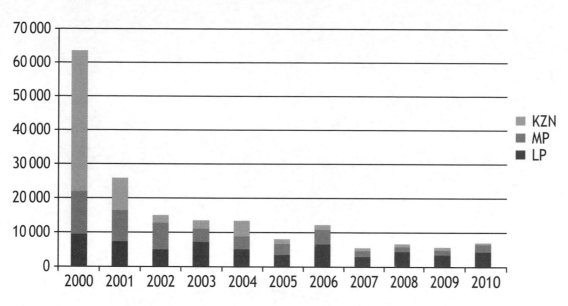

Key

KZN = KwaZulu Natal Province   **MP** = Mpumalanga Province   **LP** = Limpopo Province

Study Diagram Q14.

(a) **Describe, in detail,** the changes in malaria cases in the selected provinces of South Africa.   4

(b) For malaria, kwashiorkor, pneumonia or cholera, **explain** methods used by countries to control the spread of the disease.   6

Total marks 10

[END OF MODEL PAPER]

National
Qualifications
2014

X733/75/01

Geography

THURSDAY, 29 MAY
9:00 AM — 10:30 AM

Total marks — 60

SECTION 1   PHYSICAL ENVIRONMENTS — 20 marks

Attempt EITHER question 1 **OR** question 2. **ALSO** attempt questions 3 and 4.

SECTION 2 — HUMAN ENVIRONMENTS — 20 marks

Attempt questions 5, 6, 7 and 8

SECTION 3 — GLOBAL ISSUES — 20 marks

Attempt any TWO of the following

Question   9 — Climate Change
Question 10 — Impact of Human Activity on the Natural Environment
Question 11 — Environmental Hazards
Question 12 — Trade and Globalisation
Question 13 — Tourism
Question 14 — Health

Write your answers clearly in the answer booklet provided. In the answer booklet you must clearly identify the question number you are attempting.

Use **blue** or **black** ink.

Credit will always be given for appropriately labelled sketch maps and diagrams.

Before leaving the examination room you must give your answer booklet to the Invigilator; if you do not, you may lose all the marks for this paper.

Extract No 2072/15 & 16

1:50 000 Scale
Landranger Series

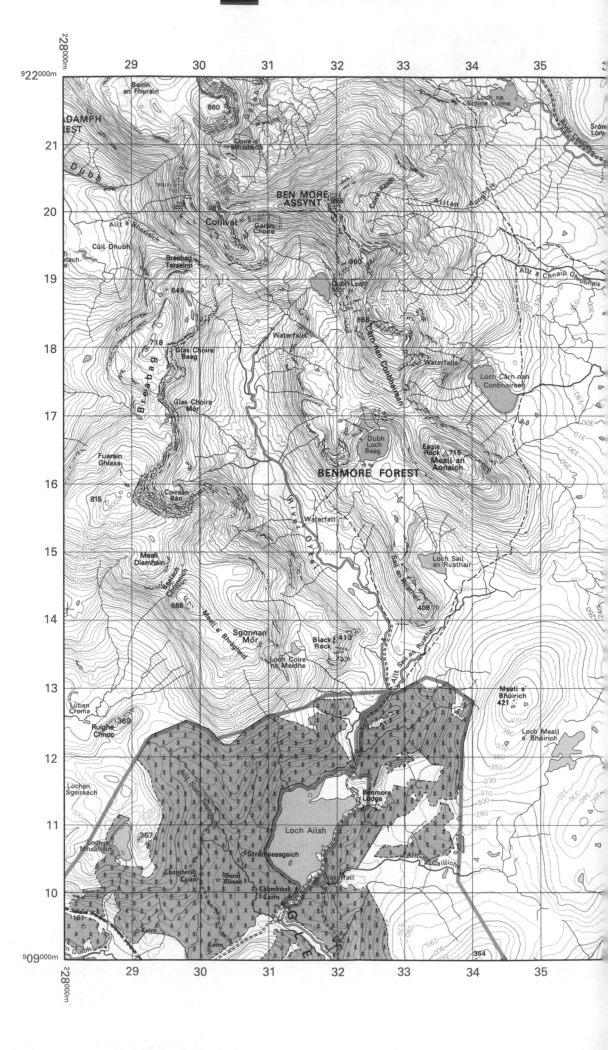

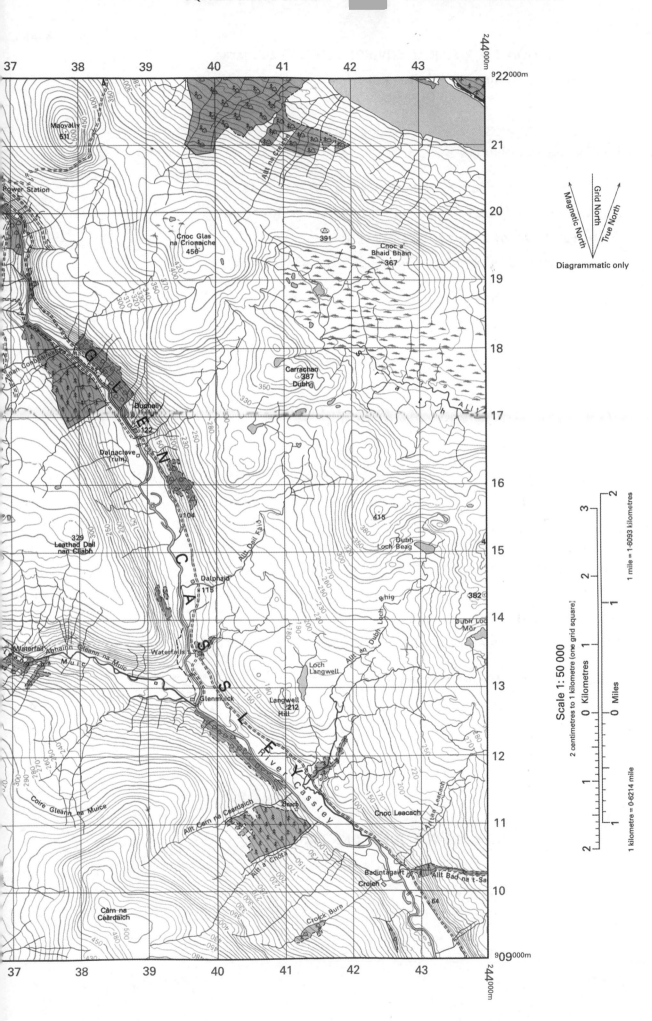

Scale 1: 50 000

2 centimetres to 1 kilometre (one grid square)

1 mile = 1·6093 kilometres

1 kilometre = 0·6214 mile

Diagrammatic only

Magnetic North   Grid North   True North

MARKS

SECTION 1 — PHYSICAL ENVIRONMENTS — 20 marks
Attempt EITHER Question 1 or Question 2
AND Questions 3 and 4

Question 1 — Glaciated Uplands

(a) Study the Ordnance Survey Map Extract (No 2072/15 & 16).

Match these glaciated features with the correct grid references

Features: **U-shaped valley; corrie; arête**

Choose from grid references: 354135, 309201, 323143, 326168.   **3**

(b) **Explain** the formation of a corrie.

You may use a diagram(s) in your answer.   **4**

Total marks   **7**

NOW ANSWER QUESTIONS 3 AND 4

DO NOT ANSWER THIS QUESTION IF YOU HAVE ALREADY ANSWERED QUESTION 1

Question 2 — River and Valleys

(a) Study the Ordnance Survey Map Extract (No 2072/15 & 16).

Match these river features with the correct grid references

Features: **ox-bow lake; meander; V-shaped valley**

Choose from grid references: 389151, 297207, 427099, 423107.   **3**

(b) **Explain** the formation of a waterfall.

You may use a diagram(s) in your answer.   **4**

Total marks   **7**

NOW ANSWER QUESTIONS 3 AND 4

MARKS

Question 3

Diagram Q3 — Land Uses

Look at Diagram Q3 and the whole of the OS Map extract.

For **one** of the land uses shown give the advantages of this area for your chosen land use. You must use map evidence.

5

[Turn over

Question 4

Diagram Q4A — Air Masses affecting the British Isles.

(a) Look at Diagram Q4A.

Describe how a prolonged spell with a **tropical continental** air mass in **summer** would affect the people of the British Isles.

3

MARKS

Diagram Q4B — Synoptic Chart for March 2012

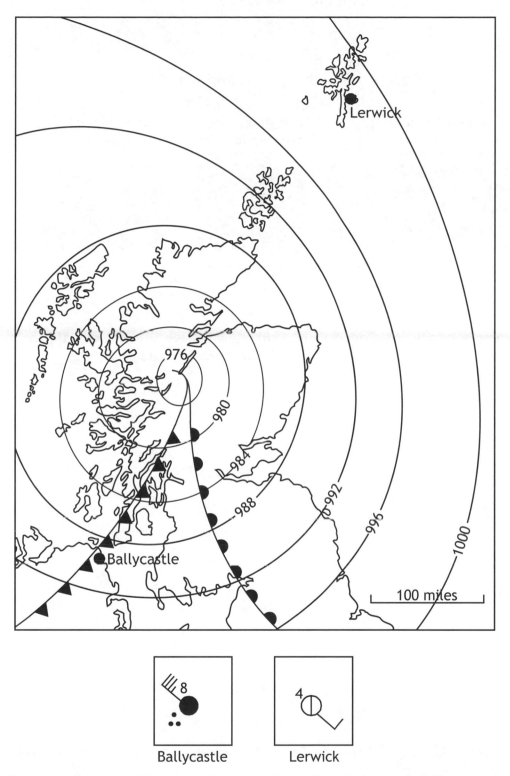

Ballycastle   Lerwick

(b) Study the Synoptic Chart and the synoptic symbols in Diagram Q4B.

  Give reasons for the **differences** in the weather conditions between Ballycastle
  and Lerwick.   5

Total marks   8

MARKS

SECTION 2 — HUMAN ENVIRONMENTS — 20 marks

Attempt Questions 5, 6, 7 and 8

Question 5

Diagram Q5 — Changes in Glasgow's CBD

Expensive Designer Stores	Old Warehouses Converted Into Flats	Indoor Shopping Malls

Look at Diagram Q5.

For Glasgow, or any other developed world city you have studied, **explain** the main changes which have taken place in the CBD over recent years.

5

MARKS

Question 6

Diagram Q6 — Age Groups in China 1982 to 2050 (Projected)

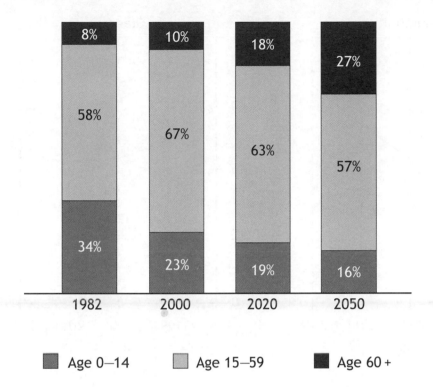

Study Diagram Q6.

(a) **Describe**, **in detail**, the changes in China's age groups between 1982 and 2050 (projected).   4

(b) For China, or other countries you have studied, **describe** methods which have been used to reduce population growth.   4

Total marks   8

[Turn over

MARKS

Question 7

Diagram Q7 — Population Trends in Mumbai and Glasgow

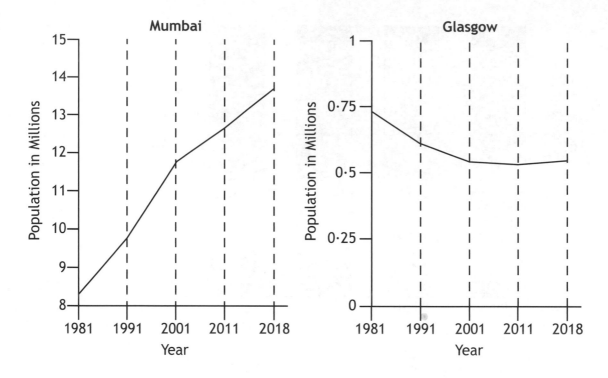

Study Diagram Q7.

Describe, **in detail**, differences in population trends between Mumbai and Glasgow.   3

Question 8

Diagram Q8 — Developments in Farming

GM Crops

Biofuel

Increased Technology

Pesticides/Fertilisers

Look at Diagram Q8.

Explain how recent developments in agriculture in developing countries are helping farmers.

4

[Turn over

SECTION 3 — GLOBAL ISSUES — 20 marks

Attempt any TWO questions

MARKS

Question 9 — Climate Change

Diagram Q9A — Areas at greater Risk from Climate Change

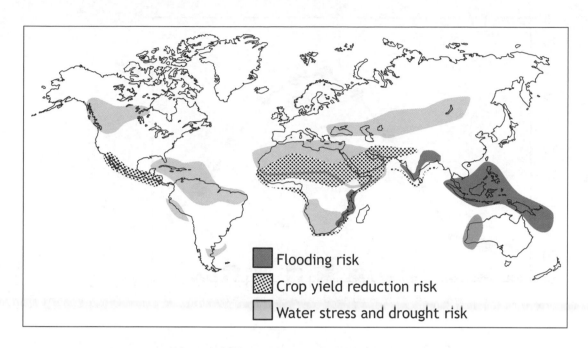

(a) Study Diagram Q9A.

Describe, **in detail**, the locations which are at greater risk from climate change.   4

Diagram Q9B — Evidence of Climate Change

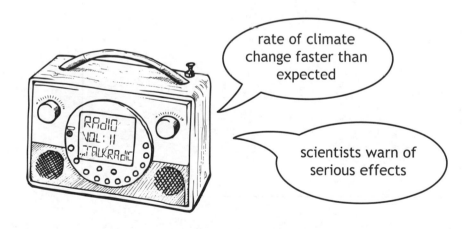

(b) Look at Diagram Q9B.

Explain, **in detail**, the physical **and** human causes of global climate change.   6

Total marks   10

[Turn over

MARKS

QUESTION 10 — Impact of Human Activity on the Natural Environment

Diagram Q10A — Barrow Alaska

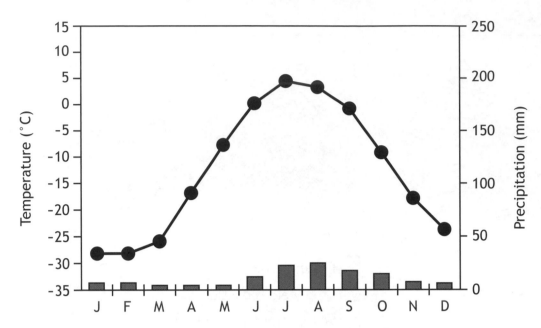

Diagram Q10B — Location of Tundra Climatic Area

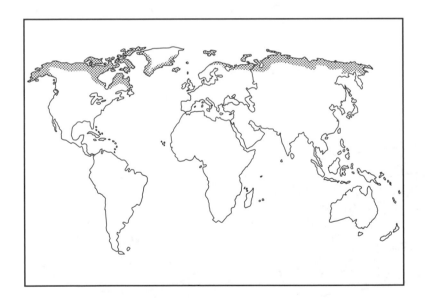

(a) Study Diagrams Q10A and Q10B.

Describe, **in detail**, the main features of the Tundra climate.   4

MARKS

Diagram Q10C — Impact of Human Activities in the Tundra

Standard of living

Future opportunities

World Wildlife Fund concerns

Wildlife habitat

Economic expansion

(b) Look at Diagram Q10C.

Describe the advantages **and** disadvantages brought about by human activities in the Tundra.

You may refer to an area you have studied in your answer.   6

Total marks   10

[Turn over

Question 11 — Environmental Hazards

Diagram Q11A — Distribution of Volcanoes

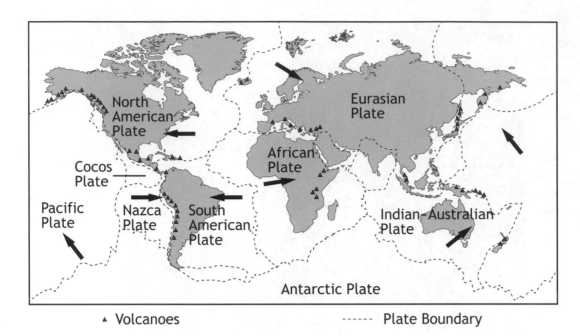

▲ Volcanoes   ----- Plate Boundary

(a) Study Diagram Q11A.

Describe, in detail, the distribution of the world's volcanoes.   4

MARKS

Diagram Q11B — Japanese Earthquake, 2011

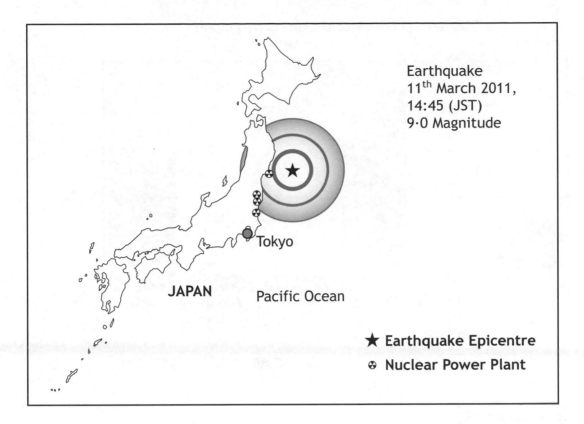

Earthquake
11th March 2011,
14:45 (JST)
9·0 Magnitude

Tokyo

JAPAN   Pacific Ocean

★ **Earthquake Epicentre**
☢ **Nuclear Power Plant**

(b) Look at Diagram Q11B.

For the Japanese earthquake, or any other named earthquake you have
studied, **describe**, **in detail**, the impact on people **and** the landscape.   6

Total marks   10

[Turn over

Question 12 — Trade and Globalisation

Diagram Q12A — World Exports by Region

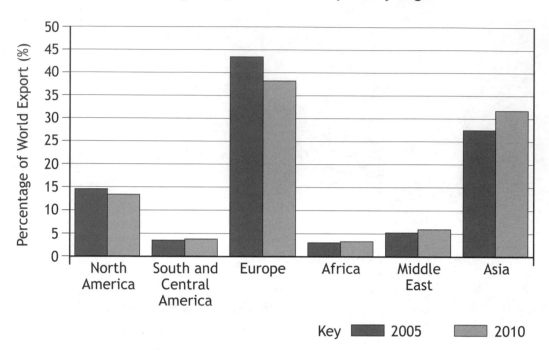

(a) Study Diagram Q12A.

   Describe, **in detail**, the change in World Exports from 2005 to 2010.   **4**

MARKS

Diagram Q12B — Collecting Fairtrade Coffee Beans

(b) Look at Diagram Q12B.

Explain how buying Fairtrade products helps people in the developing world.   6

Total marks   10

[Turn over

MARKS

Question 13 — Tourism

Diagram Q13A — Top Ten World Tourist Destinations (millions of visitors per year)

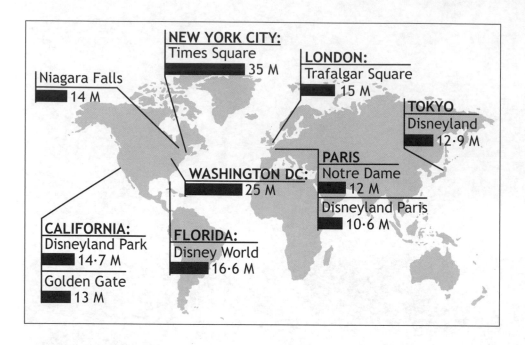

(a) Study Diagram Q13A.

Describe, **in detail**, the distribution of the top ten world tourist destinations. 4

MARKS

Diagram Q13B — Mass Tourism on an Italian Beach

jan kranendonk/shutterstock.com

(b) Look at Diagram Q13B.

Describe the effects of mass tourism on people **and** the environment.   6

Total marks   10

[Turn over

MARKS

Question 14 — Health

Diagram Q14A — Worldwide Child Deaths under 5 Years (per 1,000 live births)

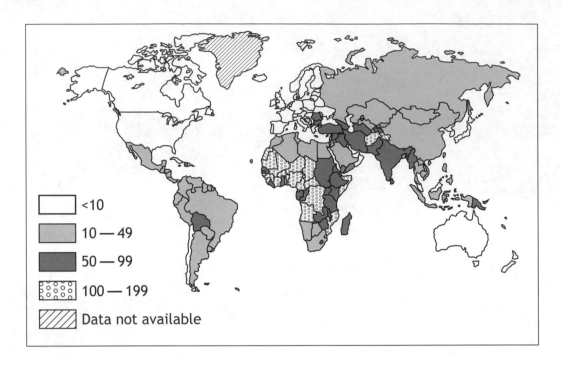

(a) Study Diagram Q14A.

Describe, **in detail**, the distribution of worldwide child deaths under the age of 5.

4

MARKS

Diagram Q14B — World AIDS Day Report 2012

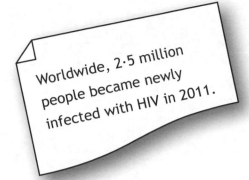

Worldwide, 2·5 million people became newly infected with HIV in 2011.

The number of people accessing HIV treatment increased by 63% from 2009 to 2011.

(b) Look at Diagram Q14B.

Explain methods used to limit the spread of AIDS in developed **and** developing countries.

6

Total marks   10

[END OF QUESTION PAPER]

[BLANK PAGE]

National Qualifications 2015

X733/75/11   **Geography**

THURSDAY, 21 MAY
9:00 AM – 10:45 AM

Total marks — 60

SECTION 1 — PHYSICAL ENVIRONMENTS — 20 marks

Attempt EITHER question 1 **OR** question 2. **ALSO** attempt questions 3, 4 and 5.

SECTION 2 — HUMAN ENVIRONMENTS — 20 marks

Attempt questions 6, 7 and 8

SECTION 3 — GLOBAL ISSUES — 20 marks

Attempt any TWO of the following

Question   9 — Climate Change
Question 10 — Impact of Human Activity on the Natural Environment
Question 11 — Environmental Hazards
Question 12 — Trade and Globalisation
Question 13 — Tourism
Question 14 — Health

Credit will always be given for appropriately labelled sketch maps and diagrams.

Write your answers clearly in the answer booklet provided. In the answer booklet you must clearly identify the question number you are attempting.

Use **blue** or **black** ink.

Before leaving the examination room you must give your answer booklet to the Invigilator; if you do not, you may lose all the marks for this paper.

MARKS

SECTION 1 — PHYSICAL ENVIRONMENTS — 20 marks
Attempt EITHER Question 1 or Question 2
AND Questions 3, 4 and 5

Question 1 — Coastal Landscapes

Study the Ordnance Survey Map Extract (Item A) of the Salcombe area.

(a) Match these grid references with the correct coastal features
Grid references: **766356, 674398, 690382**
Choose from features: cliff; headland; bay; stack.

3

(b) **Explain** the formation of **one** of the coastal features listed in part (a).
You may use a diagram(s) in your answer.

4

NOW ANSWER QUESTIONS 3, 4 AND 5

DO NOT ANSWER THIS QUESTION IF YOU HAVE ALREADY ANSWERED QUESTION 1

Question 2 — Rivers and Valleys

Study the Ordnance Survey Map Extract (Item A) of the Salcombe area.

(a) Match these grid references with the correct river features
Grid references: **708473, 713410, 684466**
Choose from features: levée; meander; v-shaped valley; waterfall.

3

(b) **Explain** the formation of **one** of the river features listed in part (a).
You may use a diagram(s) in your answer.

4

NOW ANSWER QUESTIONS 3, 4 AND 5

Question 3

Diagram Q3: Quote from a Local Landowner

> "This area has the potential for a variety of different land uses, including farming, forestry, recreation/tourism, water storage/supply, industry and renewable energy."

Study Diagram Q3 and the Ordnance Survey Map Extract (Item A) of the Salcombe area.

Choose **two** different land uses shown in Diagram Q3.

Using map evidence, **explain** how the area shown on the map extract is suitable for your chosen land uses.

5

Question 4

Diagram Q4: Selected Land Uses

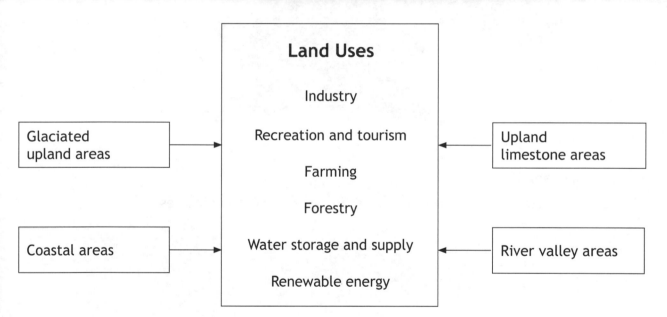

Look at Diagram Q4 above.

For a named area you have studied, **explain**, **in detail**, ways in which **two** different land uses may be in conflict with each other.

4

[Turn over

MARKS

Question 5

Diagram Q5: Average UK Temperatures in July

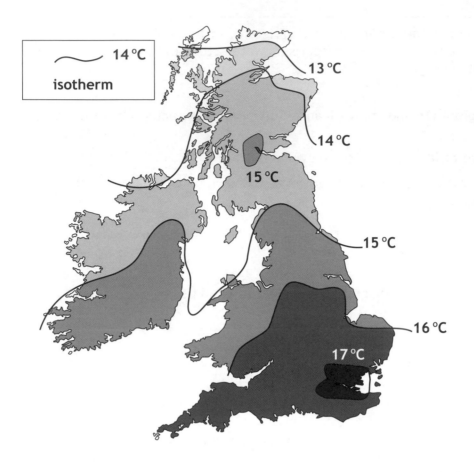

Look at Diagram Q5.

Explain the factors which cause differences in average UK temperatures.   4

MARKS

SECTION 2 — HUMAN ENVIRONMENTS — 20 marks
Attempt Questions 6, 7 and 8

Question 6

Diagram Q6

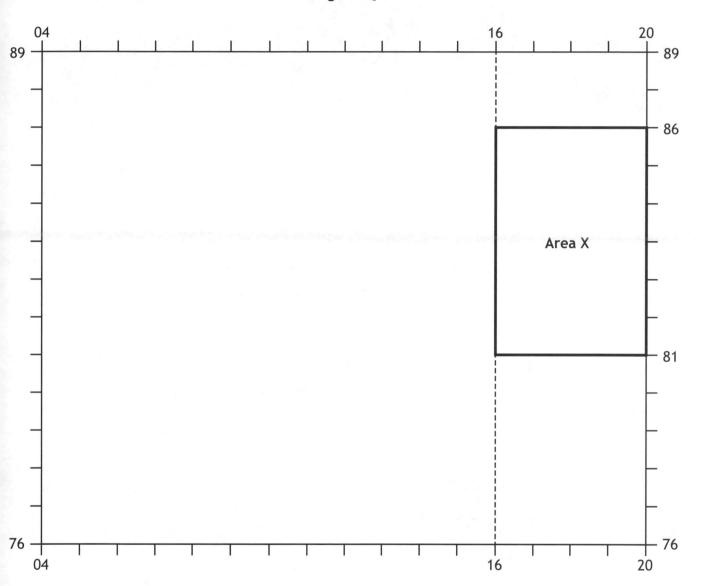

Study the Ordnance Survey Map Extract (Item B) of the Birmingham area and Diagram Q6 above.

(a) Give map evidence to show that part of the Central Business District (CBD) of Birmingham is found in grid square 0786.   **3**

(b) Find Area X on Diagram Q6 and the map extract (Item B).

Birmingham Airport, a golf course, a business park and a housing area are found in Area X on the rural/urban fringe of Birmingham. Using map evidence **explain** why such developments are found there.   **5**

Question 7

Diagram Q7: Births in Scotland 1901–2011

Look at Diagram Q7

Give reasons why the birth rate has decreased in developed countries such as Scotland.

6

MARKS

Question 8

Diagram Q8: Shanty Town Improvements in Brazil

Look at Diagram Q8

For a named city in the developing world **describe**, **in detail**, measures taken to improve conditions in shanty towns.

6

[Turn over

MARKS

SECTION 3 — GLOBAL ISSUES — 20 marks

Attempt any TWO questions

MARKS

Question 9 — Climate Change

Diagram Q9: Area of Arctic Sea Ice (1979–2013)

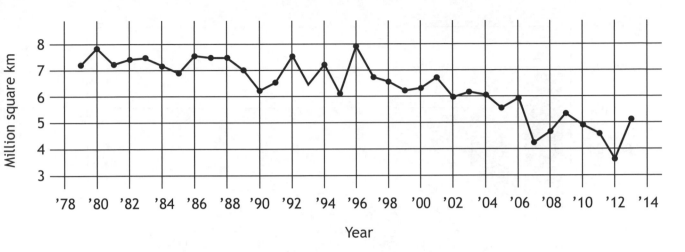

Study Diagram Q9

(a) **Describe**, **in detail**, the changes in the area of Arctic Sea ice.   4

(b) Melting sea ice is one effect of climate change.

   Explain some other effects of climate change.   6

[Turn over

MARKS

Question 10 — Impact of Human Activity on the Natural Environment

Diagram Q10A: Deforestation in Peru 2004—2012

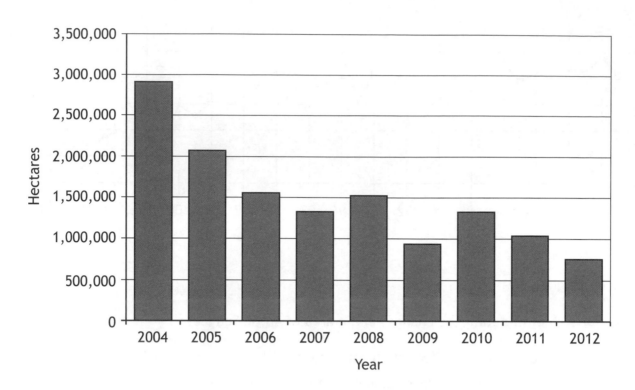

(a) Study Diagram Q10A.

Describe, in detail, the changes in deforestation in Peru from 2004 to 2012.   4

Diagram Q10B: Human Activity in the Tundra and Equatorial regions.

Oil pipeline in the Tundra   Cattle ranching in the Rainforest

(b) Look at Diagram Q10B.

For a named area you have studied, **explain** the impact of recent human activity on people and the environment.   6

MARKS

Question 11 — Environmental Hazards

Diagram Q11A: Earthquake Threatened Cities

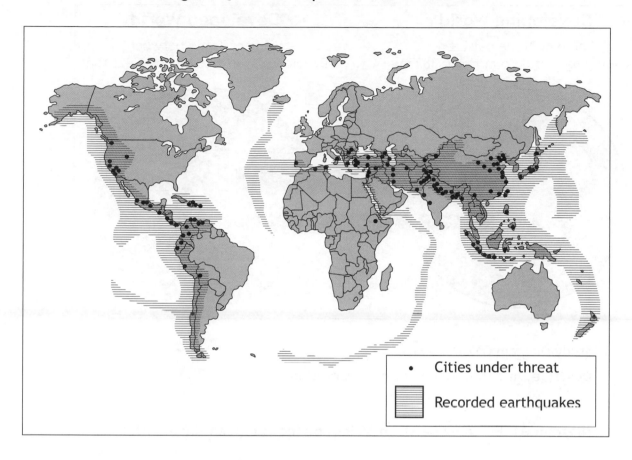

Cities under threat

Recorded earthquakes

(a) Study Diagram Q11A.

Describe, **in detail**, the distribution of cities most threatened by earthquakes.   4

Diagram Q11B

(b) Look at Diagram Q11B.

Explain, **in detail**, the strategies used to reduce the impact of an earthquake.

You must refer to named examples you have studied in your answer.   6

Question 12 — Trade and Globalisation   MARKS

Diagram Q12A: Pattern of World Trade

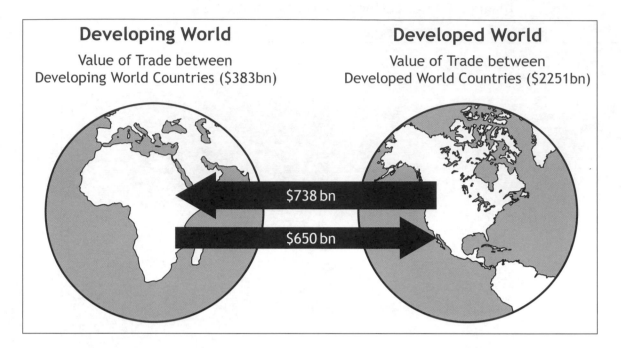

Developing World
Value of Trade between
Developing World Countries ($383bn)

Developed World
Value of Trade between
Developed World Countries ($2251bn)

$738 bn

$650 bn

(a) Study Diagram Q12A.

  Describe, **in detail**, the pattern of world trade.   4

Diagram Q12B: Trade Between Africa and the European Union (EU)

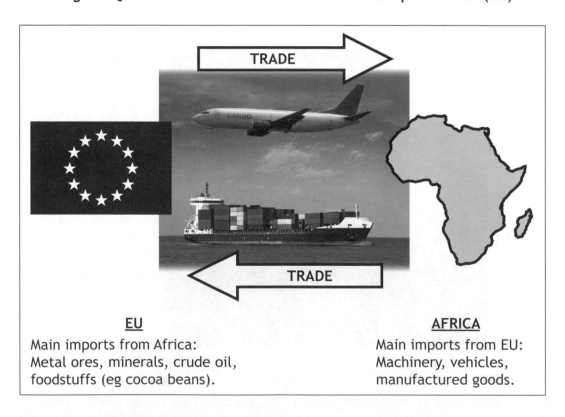

TRADE

CARGO

TRADE

EU
Main imports from Africa:
Metal ores, minerals, crude oil,
foodstuffs (eg cocoa beans).

AFRICA
Main imports from EU:
Machinery, vehicles,
manufactured goods.

(b) Look at Diagram Q12B.

  Referring to example(s) you have studied, **describe** the impact of world trade
  on people and the environment.   6

MARKS

Question 13 — Tourism

Diagram Q13A: Global Visitor Numbers: 1995–2013

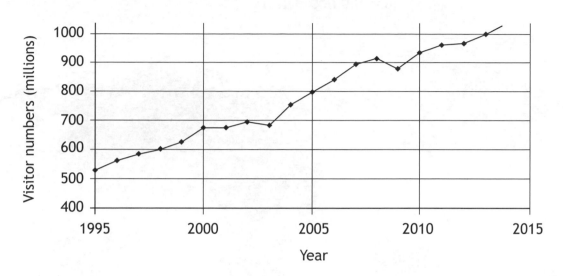

(a) Study Diagram Q13A.

Describe, **in detail**, the changes in global visitor numbers since 1995.   4

Diagram Q13B: Quote from Rainforest Community Leader

> "ECO-TOURISM has helped us to support environmental protection and improve the well-being of our people all year round."

(b) Look at Diagram Q13B.

For a named tourist area you have studied, **describe**, **in detail** the impact of eco-tourism on people **and** the environment.   6

[Turn over for Question 14 on *Page fourteen*

MARKS

Question 14 — Health

**Diagram Q14A: Worldwide Male Deaths from Heart Disease in 2011
(per 100,000 Males)**

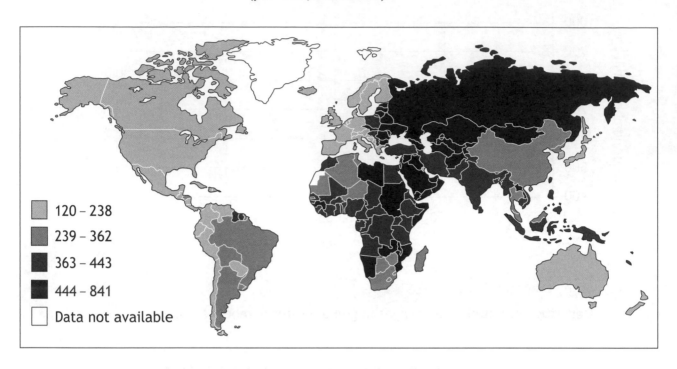

120 – 238
239 – 362
363 – 443
444 – 841
Data not available

(a) Study Diagram Q14A.

   Describe, in detail, the distribution of male deaths from heart disease.   4

Diagram Q14B: Selected Developing World Diseases

- **Malaria**
- **Cholera**
- **Kwashiorkor**
- **Pneumonia**

(b) Choose **one** disease from Diagram Q14B above.

   For the disease you have chosen, **explain** the methods used to control it.   6

[END OF QUESTION PAPER]

[BLANK PAGE]

DO NOT WRITE ON THIS PAGE

[BLANK PAGE]

DO NOT WRITE ON THIS PAGE

X733/75/21

**Geography
Ordnance Survey Map
Item A**

THURSDAY, 21 MAY

9:00 AM – 10:45 AM

The colours used in the printing of these map extracts are indicated in the four little boxes at the top of the map extract. Each box should contain a colour; if any does not, the map is incomplete and should be returned to the Invigilator.

Extract No 2143/202

1:50 000 Scale
Landranger Series

ROADS AND PATHS

Not necessarily rights of way

Junction number
Service area   Elevated
S   M1
Unfenced
A 470   Dual carriageway   Motorway (dual carriageway)
A 493   Footbridge   Primary Route (recommended through route)
    Main road
    Road under construction
B 4518   Secondary road
A 855   Bridge   B 885   Narrow road with passing places
    Road generally more than 4m wide
    Road generally less than 4m wide
    Path / Other road, drive or track
    Gradient: steeper than 20% (1 in 5),
    14% to 20% (1 in 7 to 1 in 5)
    Gates, Road tunnel
Ferry P   Ferry V   Ferry (passenger), Ferry (vehicle)

RAILWAYS

Track multiple or single   Bridges, footbridge
Track under construction   Level crossing   LC
Siding   Viaduct, embankment
Tunnel, cuttings   Station, (a) principal   a
Light rapid transit system,   Light rapid transit system
narrow gauge or tramway   station

WATER FEATURES

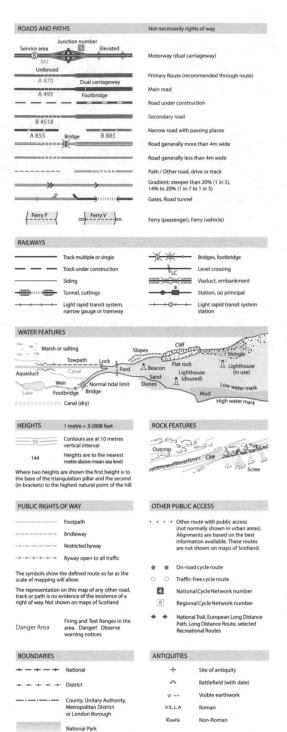

Marsh or salting   Slopes   Cliff
Towpath   Lock   Shingle
Aqueduct   Canal   Ford   Flat rock   Lighthouse
Weir   Beacon   Lighthouse   (in use)
Lake   Footbridge   Normal tidal limit   Sand   (disused)
    Bridge   Dunes   Low water mark
Canal (dry)   Mud   High water mark

HEIGHTS   1 metre = 3·2808 feet

50   Contours are at 10 metres
    vertical interval
·144   Heights are to the nearest
    metre above mean sea level
Where two heights are shown the first height is to
the base of the triangulation pillar and the second
(in brackets) to the highest natural point of the hill

ROCK FEATURES

Outcrop   Cliff   650
    600   Scree

PUBLIC RIGHTS OF WAY

Footpath
Bridleway
Restricted byway
Byway open to all traffic

The symbols show the defined route so far as the
scale of mapping will allow.

The representation on this map of any other road,
track or path is no evidence of the existence of a
right of way. Not shown on maps of Scotland

Danger Area   Firing and Test Ranges in the
area. Danger! Observe
warning notices.

OTHER PUBLIC ACCESS

• • • •   Other route with public access
(not normally shown in urban areas).
Alignments are based on the best
information available. These routes
are not shown on maps of Scotland.

●   ●   On-road cycle route
○   ○   Traffic-free cycle route
4   National Cycle Network number
8   Regional Cycle Network number
◆   ◆   National Trail, European Long Distance
Path, Long Distance Route, selected
Recreational Routes

BOUNDARIES

National
District
County, Unitary Authority,
Metropolitan District
or London Borough
National Park

ANTIQUITIES

+   Site of antiquity
⚔   Battlefield (with date)
☆ ····   Visible earthwork
VILLA   Roman
Castle   Non-Roman

TOURIST INFORMATION

⚐ 🚐 ⚐   Camp site / caravan site
✿   Garden
⛳   Golf course or links
ℹ ℹ   Information centre (all year / seasonal)
🦌   Nature reserve
P PaR   Parking, Park and ride (all year / seasonal)
✗   Picnic site
⊛   Recreation / leisure / sports centre
    Selected places of tourist interest
✆ ✆   Telephone, public / roadside assistance
☀   Viewpoint
V   Visitor centre
!   Walks / Trails
◉   World Heritage site or area
▲   Youth hostel

LAND FEATURES

x——x   Electricity transmission line
(pylons shown at standard spacing)
> - - > - >   Pipe line
(arrow indicates direction of flow)
b ruin
    Buildings
    Important building (selected)
    Bus or coach station
    Current or   with tower
    former place
    of worship   with spire, minaret or dome
+   Place of worship
⊘   Glass structure
H   Heliport
△   Triangulation pillar
T   Mast
Ⴕ   Wind pump, wind turbine
Ⴕ   Windmill with or without sails
+   Graticule intersection at 5' intervals
    Cutting, embankment
    Landfill site or slag/spoil heap
    Coniferous wood
    Non-coniferous wood
    Mixed wood
    Orchard
    Park or ornamental ground
    Forestry Commission land
    National Trust (always open / limited access,
    observe local signs)
    National Trust for Scotland (always open /
    limited access, observe local signs)

ABBREVIATIONS

Br   Bridge   MS   Milestone
Cemy   Cemetery   Mus   Museum
CG   Cattle grid   P   Post office
CH   Clubhouse   PC   Public convenience (in rural areas)
Fm   Farm   PH   Public house
Ho   House   Sch   School
MP   Milepost   TH   Town Hall, Guildhall or equivalent

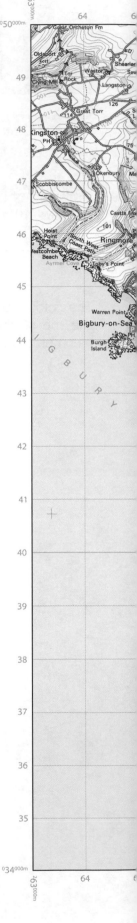

Magnetic North

Grid North

True North

Diagrammatic only

Scale 1 : 50 000

2 centimetres to 1 kilometre (one grid square)

Kilometres

Miles

1 mile = 1·6093 kilometres

1 kilometre = 0·6214 mile

[BLANK PAGE]

DO NOT WRITE ON THIS PAGE

National Qualifications 2015

X733/75/31

THURSDAY, 21 MAY
9:00 AM – 10:45 AM

Geography
Ordnance Survey Map
Item B

The colours used in the printing of these map extracts are indicated in the four little boxes at the top of the map extract. Each box should contain a colour; if any does not, the map is incomplete and should be returned to the Invigilator.

Extract No 2142/139

ROADS AND PATHS

Not necessarily rights of way

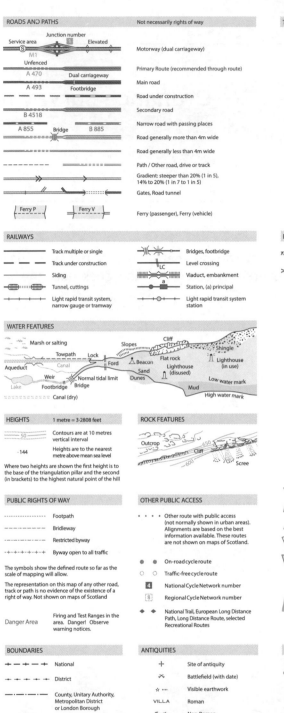

Motorway (dual carriageway)	
Primary Route (recommended through route)	A 470
Main road	A 493
Road under construction	
Secondary road	B 4518
Narrow road with passing places	A 855
Road generally more than 4m wide	B 885
Road generally less than 4m wide	
Path / Other road, drive or track	
Gradient: steeper than 20% (1 in 5), 14% to 20% (1 in 7 to 1 in 5)	
Gates, Road tunnel	
Ferry (passenger), Ferry (vehicle)	Ferry P, Ferry V

RAILWAYS

Track multiple or single		Bridges, footbridge	
Track under construction		Level crossing	LC
Siding		Viaduct, embankment	
Tunnel, cuttings		Station, (a) principal	a
Light rapid transit system, narrow gauge or tramway		Light rapid transit system station	

WATER FEATURES

Marsh or salting	Cliff
Towpath, Lock	Slopes
Aqueduct, Canal	Ford, Beacon, Flat rock, Shingle
Lake, Weir, Footbridge, Bridge	Sand, Dunes, Lighthouse (disused), Lighthouse (in use)
Normal tidal limit	Low water mark
Canal (dry)	Mud, High water mark

HEIGHTS

1 metre = 3·2808 feet

Contours are at 10 metres vertical interval — 50

·144   Heights are to the nearest metre above mean sea level

Where two heights are shown the first height is to the base of the triangulation pillar and the second (in brackets) to the highest natural point of the hill

ROCK FEATURES

Outcrop, Cliff, Scree —650 —600

PUBLIC RIGHTS OF WAY

Footpath	
Bridleway	
Restricted byway	
Byway open to all traffic	

The symbols show the defined route so far as the scale of mapping will allow.

The representation on this map of any other road, track or path is no evidence of the existence of a right of way. Not shown on maps of Scotland

Danger Area   Firing and Test Ranges in the area. Danger! Observe warning notices.

OTHER PUBLIC ACCESS

Other route with public access (not normally shown in urban areas). Alignments are based on the best information available. These routes are not shown on maps of Scotland.	
On-road cycle route	
Traffic-free cycle route	
National Cycle Network number	4
Regional Cycle Network number	5
National Trail, European Long Distance Path, Long Distance Route, selected Recreational Routes	

BOUNDARIES

National	
District	
County, Unitary Authority, Metropolitan District or London Borough	
National Park	

ANTIQUITIES

+	Site of antiquity
✗	Battlefield (with date)
☆ ····	Visible earthwork
VILLA	Roman
Castle	Non-Roman

TOURIST INFORMATION

Camp site / caravan site	
Garden	
Golf course or links	
Information centre (all year / seasonal)	
Nature reserve	
Parking, Park and ride (all year / seasonal)	P, P&R
Picnic site	
Recreation / leisure / sports centre	
Selected places of tourist interest	
Telephone, public / roadside assistance	
Viewpoint	
Visitor centre	V
Walks / Trails	
World Heritage site or area	
Youth hostel	▲

LAND FEATURES

Electricity transmission line (pylons shown at standard spacing)	
Pipe line (arrow indicates direction of flow)	
Buildings	ruin
Important building (selected)	
Bus or coach station	
Current or former place of worship	with tower / with spire, minaret or dome
Place of worship	+
Glass structure	
Heliport	H
Triangulation pillar	△
Mast	
Wind pump, wind turbine	
Windmill with or without sails	
Graticule intersection at 5' intervals	
Cutting, embankment	
Landfill site or slag/spoil heap	
Coniferous wood	
Non-coniferous wood	
Mixed wood	
Orchard	
Park or ornamental ground	
Forestry Commission land	
National Trust (always open / limited access, observe local signs)	
National Trust for Scotland (always open / limited access, observe local signs)	

ABBREVIATIONS

Br	Bridge	MS	Milestone
Cemy	Cemetery	Mus	Museum
CG	Cattle grid	P	Post office
CH	Clubhouse	PC	Public convenience (in rural areas)
Fm	Farm	PH	Public house
Ho	House	Sch	School
MP	Milepost	TH	Town Hall, Guildhall or equivalent

1:50 000 Scale
Landranger Series

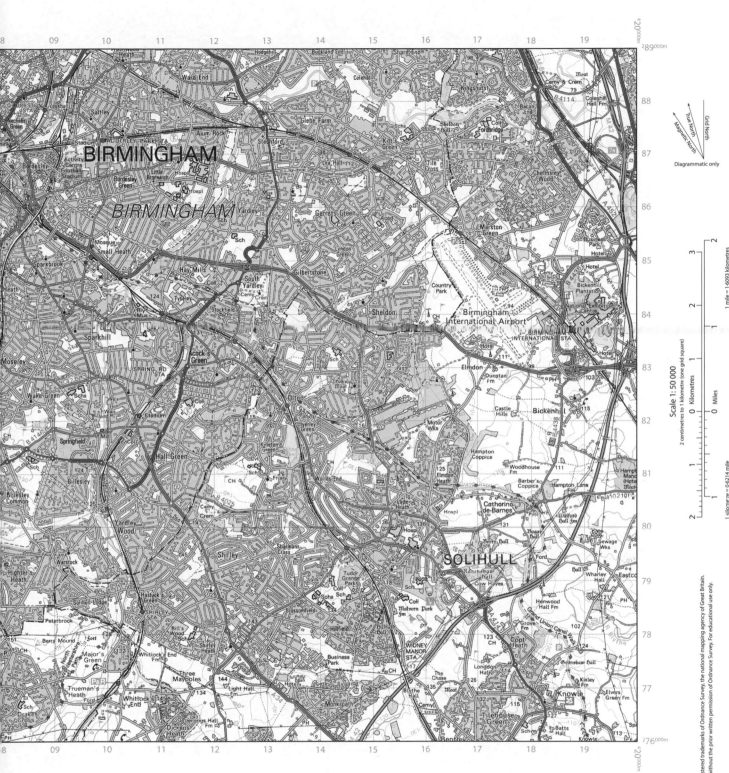

Scale 1:50 000

Extract produced by Ordnance Survey 2014.
© Crown copyright 2011. All rights reserved.

Ordnance Survey, OS, the OS Symbol and Landranger are registered trademarks of Ordnance Survey, the national mapping agency of Great Britain.
Reproduction in whole or in part by any means is prohibited without the prior written permission of Ordnance Survey. For educational use only.

[BLANK PAGE]

DO NOT WRITE ON THIS PAGE

SQA AND HODDER GIBSON NATIONAL 5 GEOGRAPHY 2015

General Marking Principles for National 5 Geography

Questions that ask candidates to *Describe* . . . (4–6 marks)

Candidates must make a number of relevant, factual points. These should be key points. The points do not need to be in any particular order. Candidates may provide a number of straightforward points or a smaller number of developed points, or a combination of these.

Up to the total mark allocation for this question:

- **One mark** should be given for each accurate relevant point.
- **Further marks** should be given for development and exemplification.

Question: Describe, in detail, the effects of two of the factors shown. (Modern factors affecting farming).

Example:

New technology has led to increased crop yields *(one mark)*, leading to better profits for some farmers *(a second mark for development)*.

Questions that ask candidates to *Explain* . . . (4–6 marks)

Candidates must make a number of points that make the process/situation plain or clear, for example by showing connections between factors or causal relationships between events or processes. These should be key reasons and may include theoretical ideas. There is no need for any prioritising of these reasons. Candidates may provide a number of straightforward reasons or a smaller number of developed reasons, or a combination of these. The use of the command word 'explain' will generally be used when candidates are required to demonstrate knowledge and understanding. However, depending on the context of the question the command words 'give reasons' may be substituted.

If candidates produce fully labelled diagrams they may be awarded up to full marks if the diagrams are sufficiently accurate and detailed.

Up to the total mark allocation for this question:

- **One mark** should be given for each accurate relevant point.
- **Further marks** should be given for developed explanations.

Question: Explain the formation of a U-shaped valley.

Example:
A glacier moves down a main valley which it erodes *(1 mark)* by plucking, where the ice freezes on to fragments of rock and pulls them away. *(second mark for development)*.

Questions that ask candidates to *Give reasons* . . . (4–6 marks)

Candidates must make a number of points that make the process/situation plain or clear, for example by showing connections between factors or causal relationships between events or processes. These should be key reasons and may include theoretical ideas. There is no need for any prioritising of these reasons. Candidates may provide a number of straightforward reasons or a smaller number of developed reasons, or a combination of these. The use of

the command words 'give reasons' will generally be used when candidates are required to use information from sources. However, depending on the context of the question the command word 'explain' may be substituted.

Up to the total mark allocation for this question:

- **One mark** should be given for each accurate relevant point.
- **Further marks** should be given for developed reasons.

Question: Give reasons for the differences in the weather conditions between Belfast and Stockholm.

Example:
In Stockholm it is dry, but in Belfast it is wet because Stockholm is in a ridge of high pressure whereas Belfast is in a depression *(one mark)*. Belfast is close to the warm front and therefore experiencing rain *(second mark for development)*.

Questions that ask candidates to *Match* (3–4 marks)

Candidates must match two sets of variables by using their map interpretation skills.

Up to the total mark allocation for this question:

One mark should be given for each correct answer.

Question: Match the letters A to C with the correct features.

Example: A = Forestry *(1 mark)*

Questions that ask candidates to *Give map evidence* (3–4 marks)

Candidates must look for evidence on the map and make clear statements to support their answer.

Up to the total mark allocation for this question:

Question: Give map evidence to show that part of Coventry's CBD is located in grid square 3379.

Example: Many roads meet in this square *(1 mark)*.

Questions that ask candidates to *Give advantages and/or disadvantages* (4–6 marks)

Candidates must select relevant advantages or disadvantages of a proposed development and show their understanding of their significance to the proposal. Answers may give briefly explained points or a smaller number of points which are developed to warrant further marks.

Up to the total mark allocation for this question:

- **One mark** should be given for each accurate relevant point.
- **Further marks** should be given for developed points.
- **Marks** should be awarded for accurate map evidence.

Question: Give either advantages or disadvantages of this location for a shopping centre. You must use map evidence to support your answer.

Example: There are roads and motorways close by allowing the easy delivery of goods *(1 mark)* and access for customers *(1 mark for development)*, eg the A46, M6 and M69 *(1 mark)*.

NATIONAL 5 GEOGRAPHY MODEL PAPER 1

Section 1: Physical Environments

1 (a) U-shaped valley – 979976
   Corrie – 957997
   Pyramidal peak – 954976   3

  (b) 1 mark for a valid point.
     2 marks for a developed point.
     Full marks can be gained for appropriately annotated diagrams.
        If a U-shaped valley is chosen: a glacier occupies a V-shaped valley (1), the ice moves, eroding the sides and bottom of the valley (1), through plucking (1) and abrasion (1); this makes the valley sides steeper (1) and the valley deeper (1), and when the glacier retreats a steep, deep, flat-floored U-shaped valley is left (2), the original stream seems too small for the wider valley and is known as a misfit stream (1).   4

2 (a) Pot hole – 833164
   Intermittent drainage – 817155
   Limestone pavement – 814175   3

  (b) 1 mark for a valid point, 2 marks for a developed point.
     Limestone pavement:
        Limestone made from decayed remains of skeletons of sea creatures (1); laid in horizontal layers on sea bed (1); sedimentary rocks uplifted (1); overlaying rock removed by glaciation (1); cracks appear as rock dries out (1); cracks widen into grykes by chemical weathering (1); limestone dissolved by acid rainwater (1); creates clints, or upstanding blocks (1).
     Credit should be given for appropriately annotated diagrams.   4

3 1 mark for a simple point.
  2 marks for a developed point.
  Max 1 for grid references.
  Answers could include:
     Most settlement is found along the course of the river where the land is low and suitable for building (1), e.g. at 885125 (1). Transport routes such as the B970 and the railway line follow the course of the valley as it provides a natural route way (2). The roads have had to avoid the higher, steeper ground and in some places the undulating nature of the land means cuttings and embankments are needed (1) such as at 892132 (1). Where there is gently sloping lower ground (901117) farming may be arable or mixed (1), whereas on the steeper slopes and higher ground livestock grazing will occur due to the difficulty of using machinery (2). Woodland is grown on land that is too high and cold for crops, such as 905066 (1). Steep slopes are no use for settlement but can be for recreation and tourism, such as the ski centre at 899059 (1).
  Accept any other relevant point.   5

4 Max 2 marks for description of weather changes without explanation.
  Mark out of 3 if candidate has misidentified fronts but explained weather correctly.
  Answers could include:
     As the warm front approaches Norwich, air pressure will fall (1), cloud cover will increase (1) and steady rain will occur (1). Winds will be quite strong as the isobars are close together (1). The warm front will move away, and Norwich will be in the warm sector of a low-pressure system (1). Temperatures will rise, and it will be mild with occasional showers and some cloud cover (1). Winds will die down. The cold front will arrive and cloud cover will increase, with cumulonimbus clouds bringing heavy rain to the city (2). Temperatures will drop as the cold front passes over and begins to move away (1). The sky will become clear (1), the rain will stop (1) and pressure will begin to rise (1), and winds will increase (1).
  Accept any other valid point.   4

5 (a) 1 mark for a valid point.
     2 marks for a developed point.
     If glaciated uplands chosen:
     • Farming – Highlands are suitable for sheep farming as opposed to arable farming (1). Sheep can survive on steep slopes (1). Steep land is unsuited to machinery used on arable farms (1).
     • Forestry – Trees can be planted on the upper slopes where the land is unsuitable for other crops due to weather and steepness of slopes (1).
     • Recreation and tourism – Tourists are attracted to mountain scenery (1). The steep slopes can be used for skiing (1). The mountains attract hill walkers (1).
     • Water storage and supply – Valleys can be dammed to create reservoirs (1). Mountain areas usually have high rainfall (1). The underlying rock may be impermeable and therefore suitable for water catchment and storage (1).
     • Renewable energy – Highlands provide suitable sites for HEP schemes (1). Water flowing down pipelines can be used to drive turbines to create electricity (1).
     • Mountain areas are also suitable sites for wind farms because of the frequency and strength of the local winds (1).   4

Section 2: Human Environments

6 (a) 1 mark per valid point.

Answers could include:

Main roads like the A691 and the A690 converge on this square (1); there is a tourist information centre (1) and a cathedral (1), many churches (1) and a bus station (1).

Accept any other valid point.   3

(b) 1 mark per valid point.

2 marks for a developed point. Mark only differences.

Answers may include:

2745 is an area of newer housing, 2642 is an older, inner urban area (1); 2745 is mainly a residential suburban area whereas 2642 has a greater variety of land uses like industry and services, as well as housing (2). 2745 has a varied street pattern including cul-de-sacs and crescents whereas 2642 has a mainly rectangular/grid-iron pattern (1); 2745 has mainly small buildings (houses) whereas 2642 has large buildings like factories (1); 2745 has a limited amount of traffic and less noise pollution, while 2642 has many main roads, railway, bus station – more noise and pollution (2).

Accept any other valid point.   5

7 1 mark for a valid point.

2 marks for a developed point.

If no named example, mark out of 5.

If the Rio de Janeiro example was chosen, answers may include:

New roads are to be constructed to improve the transport of people and goods in the area (1); storm drains are to be built to control flooding (1); construction of piped water supply (1); construction of latrines (1). Slums will be cleared over a five-year period (1), and people are being rehoused nearby in newly built apartments (1). These are affordable accommodation (1). The estates also include schools, markets and other facilities (1).

Or any other valid point.   6

8 For full marks, two factors should be mentioned.

Award up to 4 marks for any single factor.

1 mark per valid point.

2 marks for a developed point.

If New Technology is chosen, answers may include:

Machinery increases the efficiency on a farm, enabling the farmer to plough, sow, spray etc. more quickly, covering larger areas (2). It also speeds up harvesting and results in the product being delivered to markets fresher (1) and at a higher premium (1). It also allows for a smaller work force (1) and therefore lower wage bills (1). It allows for the use of satellite technology/computers to control the application of fertilisers to particular areas of fields (1), improving yields (1) yet decreasing the cost and waste.

If Diversification is chosen, answers may include:

Farmers can obtain additional income from a variety of sources if they diversify their activities on the farm (1). They may turn old farm workers' cottages into holiday chalets (1). They may use part of the land for a golf course (1). They may earn income from sports such as quad-bike riding (1).

Or any other valid point.   6

Section 3: Global Issues

9 (a) Answers may include:

Over a 17-year period between 1979 and 1996, ice melt decreased (1), apart from 1983-4 when it increased (1). The general trend for 1997 onwards was that ice melt increased (1), reaching a high in 2012 (1) at 2% higher than the average (1).

Or any other valid point.   4

(b) 1 mark for a valid point.

2 marks for a developed point.

Answers may include:

Warmer global temperatures could change the habitats of different wildlife and marine dwellers (1). Warmer water causes some fish to die or to move to colder waters, which affects fishermen and their livelihood (1). Increased temperatures may mean more drought, leading to famine (1). In some areas, new crops can be grown because of the higher temperatures (1), which increases farmers' income (1). Rising temperatures can cause the icecaps to melt, which can result in flooding in low-lying areas like the Netherlands (1). Warmer temperatures could result in the spread of diseases like malaria into new areas (1).

Or any other valid point.   6

10 (a) Answers may include:

The lowest temperature reached in Barrow is −29 degrees whereas in Eismitte it is −27 degrees Centigrade (1). The highest temperature reached in Barrow is 5 degrees, compared with 12 degrees Centigrade in Eismitte (1). The range in temperature in Barrow is therefore 34 degrees as opposed to 39 degrees in Eismitte (1). Barrow has precipitation throughout the year, totalling 130mm, whereas there is no rainfall in Eismitte (2).

Or any other valid point.   4

(b) Answers may include:

Large areas of rainforest are needed for cattle ranching, so huge areas are burned, destroying the native wildlife and the plants (1). The land the cattle graze soon becomes infertile, so more rainforest is destroyed to allow the cattle to graze (1). This land never recovers (1). Large tracts of forest are cleared by using fires, which can get out of control and destroy far more forest than necessary (1). Large amounts of CO_2 are released, affecting the local climate as well as the global climate (1). With no trees to bind the soil together, soil erosion takes place (1). Rivers become polluted with soil and cannot be used by the local people (1). The local people are forced off the land, and their traditional way of life is under threat (1). Logging destroys the habitats of animals, causing a threat of extinction (1). Fewer animals means less food for the native people (1). Illegal loggers do not replant to replace the trees, so the rainforest does not regenerate (1).

Or any other valid point.   6

11 (a) Answers may include:

Volcanoes are found along the edges of plate boundaries (1). Many volcanoes are found along the west coast of North and South America (1) especially Alaska (1). Volcanoes are found along the coastal areas of China and Japan (1) and along the Pacific Ring of Fire (1).

Or any other valid point.　4

(b) 1 mark for a valid point.

2 marks for a developed point.

If no specific example named, mark out of 4.

Answers may include:

Tropical storms are very powerful and can cause immense damage (1). Storms can uproot trees (1) and disrupt telephone lines and electricity power lines (1). If plantation crops are destroyed, this can cause severe economic problems (1). Tidal surges flood low-lying coastal areas (1). There can be major loss of life caused by flooding, resulting in thousands of deaths (1). Landslides may occur where rainfall washes away buildings built on steep, unstable slopes (1). Flooding can block coastal escape routes and relief roads, making damaged areas unreachable for rescue services (2).

Or any other valid point.　6

12 (a) Answers may include:

The percentage share of world goods production for developed countries will steadily decrease between 2000 and 2016 (1). The percentage will have decreased from 68% to 45% in developed countries – a drop of 23% (1). China's share will steadily increase during this period from 4% to an estimated 18% (1). India's percentage has also increased, but the increase is significantly lower than that of China (1). India's share remained at 4% between 2000 and 2010 but is due to increase slightly to 7% by 2016 (1).

Or any other valid point.　4

(b) 1 mark for a valid point.

2 marks for a developed point.

Answers may include:

Developed countries have a larger share of world trade because their exports include significantly more manufactured goods than countries in the developing world (1). Developing countries tend to produce raw materials rather than manufactured goods (1). Developed countries have more industries producing a wide variety of products, such as food products, industrial machinery and electronics, which are traded with other developed countries (2). Many developed countries belong to trading alliances, such as the European Union, which help to increase the volume of trade (1). The economies of developed countries benefit from being able to purchase low-cost raw materials produced by developing countries and to sell manufactured goods back for higher profits (2). Developing countries have much less money to invest in manufacturing industries and are less able to compete with developed countries (1).

Or any other valid point.　6

13 (a) 1 mark for a valid point.

2 marks for a developed point.

Answers may include:

The overall number of tourist arrivals across the world has steadily increased throughout the period from 1990 to 2011 (1). The number has more than doubled from 1990 to 2011 (1). The highest number of arrivals in 2011 occurred in Europe, with France being the most visited country (2). The second largest number of arrivals occurred in Asia, Australia and Oceania, with China accounting for 57·6 million visitors (2). North and South America were third largest in tourist arrivals, with the USA having 62·3 million tourist visitors (2). Africa had the smallest share of tourist arrivals with only 5%, as opposed to 51·5% in Europe (2).

Or any other valid point.　4

(b) 1 mark for a valid point.

2 marks for a developed point.

For full marks, both advantages and disadvantages must be mentioned. If not, mark out of 5.

Advantages:

Mass tourism boosts the economies of developed and developing countries (1). Tourism creates jobs in a wide variety of activities, including farming (supplying food for shops/hotels etc.) and entertainment (theatres, leisure centres) (2). It provides money for the country to improve infrastructure (transport, water supplies, sewage disposal) (1) and provides income to improve services such as education and health services (1).

Disadvantages:

Tourism can create pollution, e.g. litter on beaches (1). Sea and river pollution arises from the increased use of fertilisers and pesticides on local farms (1). Can increase traffic congestion in both cities and rural areas (1). Can cause conflict in rural areas with farmers to produce more food to feed tourists (1). Employment may only be seasonal (1). Tourism can have a detrimental effect on local culture and the physical environment, for example removing forest to build more tourist facilities (1).

Or any other valid point.　6

14 (a) 1 mark for a valid point.

2 marks for a developed point.

Answers may include:

AIDS is most prevalent in countries in the developing world (1). The highest percentages of infected adults are found in areas such as Central and South Africa (1), which have 15 to 34% of the population infected (1). Infection rates are much lower in areas such as Europe and North America at 0·1 to 0·5% (1). Infection rates are also low in Australia (1). Rates are also high in parts of Asia, particularly in India and Pakistan (2).

Or any other valid point.　4

(b) (i) 1 mark for a valid point. 2 marks for a developed point.

Max 1 mark for a list.

If heart disease is chosen, answers may include:

Heart disease can be inherited from parents (1). Over-eating can lead to obesity, putting extra pressure on the heart (2). Smoking narrows the arteries and affects the lungs (1). This can lead to emphysema, putting a strain on the heart by making it work faster (1). The build-up of fatty deposits on the walls of the arteries restricts the flow of blood to the heart (1); lack of exercise raises blood pressure, affecting the efficiency of the heart (1); poor diet increases cholesterol (1); stress leads to high blood pressure (1).

Or any other valid point. **3**

(b)   (ii) Max 1 mark for a list.

Answers may include:

More people now have regular check-ups for cholesterol and blood pressure (1), allowing early intervention for at-risk patients (1). More advanced medical equipment is being invented and used (1), e.g. artificial heart valves (1). More advanced surgery is now available (1), e.g. bypass surgery (1). The government runs many campaigns to educate the public (1), e.g. stop-smoking campaigns and healthy eating (1). Advertising brands of cigarettes has been banned (1). Smoking ban in public places (1). Nicotine patches to help people give up smoking (1). People are encouraged to eat more healthily and take more exercise (1). Healthy eating is encouraged in school dining halls (1). Free and reduced membership of gyms (1).

Or any other valid point. **3**

NATIONAL 5 GEOGRAPHY MODEL PAPER 2

Section 1: Physical Environments

1 (a) Stack – 636871
Headland – 570863
Bay – 555869 **3**

(b) 1 mark for a valid point.
2 marks for a developed point.
- Caves – Caves occur where a coastline of hard rock is attacked by prolonged waves (1). The waves attack along a line of weakness such as a joint or fault in the rock (2).
- Arches – Over time, erosion of the cave may cut through the headland (1). When it reaches the other side, an arch may be formed (1).
- Stack – Continual erosion at the foot of the arch may cause the roof to collapse (1). This creates a piece of rock isolated from the headland called a stack (1).

Or any other valid point. **4**

2 (a) Confluence – 584973
Meander – 614965
River flowing south – 670967 **3**

(b) 1 mark for a valid point.
2 marks for a developed point.
A fully annotated diagram can obtain full marks.
Waterfall – Waterfalls occur in a river valley at a point where soft rock is overlain with harder rock (1). Water flowing over the hard rock begins to erode the softer rock underneath (1). At the point of erosion, a pool called a plunge pool is formed (1). As the softer rock wears away, it can no longer support the harder rock above, which collapses into the plunge pool (1).

Or any other valid point. **4**

3 1 mark per valid point.
2 marks for an extended point.
1 mark for a grid reference.
Answers may include:
There is woodland for forest walks (1) at 533903 (1). The coastline attracts tourists for the beautiful views (1). The limestone cliffs are attractive for rock-climbing (1). Oxwich Bay is attractive for its beautiful beach and sand dunes (2). Caves like Bacon Hole (560867) along the coast attract cavers (2). The sea can be used for water activities like boating, surfing and swimming (1).

Accept any other valid point. **5**

4 For full marks, both similarities and differences must be mentioned.
Otherwise mark out of 3.
Similarities:
There are few clouds in the sky (1). The weather is dry (1). Wind speeds are light and calm towards the centre of the anticyclone (1). There may be mist in the morning at both times of the year (1).
Differences:
Temperatures are high in summer and low in winter (1). During the night in summer, temperatures remain above zero (1), but in winter they will fall below zero. Warm moist air rising from the ground can form thunderstorms in summer (1). Heatwaves can occur in summer if the anticyclone remains for a period of time (1).

Or any other valid point. **4**

5 1 mark for a valid statement.
2 marks for a developed point.
If Farming/Recreation and Tourism is chosen, answers may include:

Walkers climbing over stone walls could damage them (1). Sheep may be frightened by walkers' dogs (1). Gates could be left open, allowing animals to escape (1). Tourists could be denied access to areas (1). Tourists could cause an increase in traffic congestion on small country roads (1). Tourists could drop litter which animals could eat (1). Farmers could cause some river pollution through fertilisers and pesticides draining into rivers (1).

Or any other valid point. 4

Section 2: Human Environments

6 1 mark per valid point.
2 marks for an extended point.
For full marks, both advantages and disadvantages should be mentioned. If not, mark out of 4.
Advantages:

There is flat land suitable for building on (1). There is room for expansion (1). Swansea is close by, providing a market and a labour force (2). The B4620 allows transport of goods in and out as well as easy access for customers (2). Many people will travel by car, and there is room for parking (1). The site is on the outskirts of town, so the land will be cheaper (1) and traffic congestion less likely than a location in the town centre (1).

Disadvantages:

There are electricity transmission lines taking up space in the square, limiting the available area for building (1). The Afon Llan runs through the square, which could cause flooding (1). There is only a B-class road running into the square, so it may not be able to handle the amount of traffic and the size of vehicles (2). There is already an industrial estate in the square, so traffic congestion may occur (1).

Accept any other valid point. 5

7 (a) 1 mark for a valid point.
2 marks for a developed point.
Answers may include:

The most densely populated areas include Europe, the east coast and parts of the west coast of the USA, the east coast of South America, north-west Africa, coastal India and the eastern coasts of China (2).

Moderately populated areas include the Midwest and western parts of the USA, Scandinavia and eastern and central Russia, coastal Australia and central and southern Africa (2).

Sparsely populated areas include northern Canada, central South America, central Australia and northern Africa (2).

Or any other valid point. 3

(b) 1 mark for a valid point.
2 marks for a developed point.
For full marks, both physical and human factors should be mentioned. Otherwise, mark out of 4.
Answers may include:

Physical factors such as relief, high mountains and climate, desert areas and polar regions limit population (2). Many parts of the world are inaccessible, such as tundra areas (1). People are attracted to areas of flat land such as river valleys and coastal areas because land is easier to build on (2). Soils are usually more fertile in these areas for farming (1). Areas which have natural resources such as wood, oil, coal and water attract higher numbers of people (1).

Human factors include employment opportunities; where people can find jobs; and industry (1).

Governments can offer financial incentives to attract industries which can attract employment into an area (1) and therefore people (1). If areas are easily reached through good transport links, there is likely to be more work available (1).

Or any other valid point. 6

8 1 mark for a valid point.

2 marks for a developed point.

If exports are chosen, answers may include:

The USA is more developed than Chad because it exports manufactured goods, which leads to more money coming into the country (1), whereas Chad exports mainly raw materials, which bring in much less money than manufactured goods (1). The USA will have higher levels of pay and employment than Chad, and therefore the standard of living for the population in the USA will be much higher than in Chad, where wages will be low (1).

If literacy rates are chosen, answers may include:

Literacy rates are much higher in the USA than in Chad because better standards of education are provided by the government (1). There will be more schools, colleges and universities available to educate people in the USA (1). Since 99% of people in the USA can read as opposed to 35% in Chad, people will have a better chance of obtaining employment in a wide range of jobs in the USA (1).

Or any other valid point. 6

Section 3: Global Issues

9 (a) 1 mark for a valid point.

2 marks for a developed point.

Answers may include:

Average global temperatures have steadily risen from 13·7 degrees in 1880 to a predicted 14·5 degrees in 2020 (1). Throughout that period, the temperature has fluctuated over various 20-year periods (1). From 1880 to 1920, there were four times when the temperature fell below 13·7 (1). From 1920 to 1960, the temperature reached a high of about 14·3 in 1940 (1). Again there were several times when the temperature fell during this period, such as in 1960, when it fell to 13·8 degrees (1). In the period 1960 to 2000, the highest temperature reached was in 2000, when it rose to about 14·6 degrees (1). The lowest temperature in this period was about 13·8 degrees around 1970 (1).

Or any other valid point. 4

(b) Both physical and human causes must be mentioned for full marks. Otherwise, mark out of 4.

1 mark for a valid point.

2 marks for a developed point.

Physical causes:

These include times when there were sun flares, which helped to increase temperatures (1). Gases such as methane given off from rotting vegetation in areas such as the tundra cause a natural increase in air pollution, again causing temperatures to rise (1). These changes in temperature can cause winds such as the Westerlies to change course (1). These can cause changes to the climates in different areas of the world which are affected by offshore winds (1). The temperature changes in the seas and oceans can cause changes in ocean currents (1). These can have an adverse effect on the climates of coastal areas through flooding and hurricanes (1).

Human causes:

The burning of forests to clear land for other human activities releases vast amounts of carbon dioxide into the atmosphere, which pollutes it (1). Emissions from road-transport exhaust fumes also release toxic pollution into the atmosphere (1). Industrial plants release additional gases which pollute the atmosphere (1). Some scientists have suggested that these pollutants have punctured holes in the ozone layer which protects the Earth against harmful rays from the sun (1). The use and testing of atom bombs during the 1940s, 50s and 60s released radioactive particles into the atmosphere and beneath some oceans (1). Gases known as CFCs have been released from waste sites where disused fridges have been dumped (1).

Or any other valid point. 6

10 (a) 1 mark per valid point.

2 marks for a developed point.

Answers may include:

There was a fall in the number of square miles of deforestation from 8000 to 4500 between 1988 and 1990 (1). From 1990 to 1992, deforestation increased to a peak of 11000 square miles (1). This was followed by a further drop to just under 6000 square miles by 1996 (1). From 1996 to about 2003, deforestation rose steadily to a peak of 10000 square miles (1). Since then, there has been a continual drop to just over 2000 square miles in 2010 (1).

Or any other valid point. 4

(b) 1 mark for a valid point.
2 marks for a developed point.
Answers may include:

National parks have been set up in rainforest areas to protect the land and people from the environmental effects of mining (river pollution), cattle ranching and road building (2).

In places such as Colombia, the government has returned land to local people which had been taken away from them (1). Developed countries are less inclined to fund projects which are harmful to the rainforests (1). Laws against illegal logging have been passed to reduce deforestation (1). To protect the environment, logging companies have been encouraged to replant trees in areas where trees have been removed (1). Agro-forestry schemes have been introduced whereby forests are used for small-scale farming schemes (1). These provide sustainable crop yields and protect the ways of life of local people (1).
Or any other valid point.   6

11 (a) 1 mark for a valid point.
2 marks for a developed point.
For hurricanes, answers may include:

Hurricanes occur in the Caribbean in areas such as Jamaica (1). They also occur along the south-east coast of the USA, affecting areas such as Florida (1). They occur in an area stretching from Oceania to the south-east coast of Africa (1). Typhoons are found in South-East Asia stretching across the Indian Ocean to coastal India and Bangladesh (2).
Or any other valid point.   4

(b) 1 mark for a valid point.
2 marks for a developed point.
For earthquakes, answers may include:
Scientists can predict where earthquakes will occur, but it is almost impossible to predict when (1). Seismograph readings do not give enough warning (1). Despite advanced technology, extensive damage and loss of life can happen in built-up areas such as Los Angeles or Kobe (2).
It is better to build earthquake-proof buildings in known areas where earthquakes occur (1). It is also better to plan for an earthquake by having rescue services and medical services trained and ready to deal with a disaster (1).
For volcanic eruptions, answers may include:
Tremors and gas emissions give warning to local people (1). This allows time for evacuation (1). Although the landscape can be destroyed, evacuation prevents huge loss of life (1). If a volcano is becoming active, it may expel lava bombs which will warn the population to leave the area before it erupts (1). Areas where volcanoes are active can plan to have rescue/aid measures in place to reduce the impact of an eruption, such as the destruction of buildings and roads by flowing lava (1).
For tropical storms, answers may include:
Satellite images can provide plenty of warning (1), allowing the population time to evacuate the area (1) and protect buildings (1). People can prepare to protect themselves by building storm shelters under the ground (1). Storm warnings can be given over the radio or television (1). Local people often stock up on food and water supplies to use after the storm has passed (1).
Or any other valid points.   6

12 (a) 1 mark for a valid point.
2 marks for a developed point.
Note that it is not necessary to describe the changes for all four countries for full marks.
Answers may include:

From 1972 to about 1998, the United States' percentage share of world exports fluctuated between 10 and 12%, having begun the 1970s at about 14% (2). From 1998, the USA's share fell to just over 8% (1). Germany's share was at its highest in the later part of the 1980s with a peak of just over 12% (1). Since then, it has gradually fallen to about 9% (1). Japan's share increased from 6% in about 1972 to a high of 10% in around 1985 (1). Since then, it has decreased to just under 5% in about 2008 (1). China's percentage share rose steadily from about 0·7% in 1970 to just over 3% in about 1997 (1). It then increased dramatically to over 12% by 2010 (1).
Or any other valid point.   4

(b) 1 mark for a valid point.
2 marks for a developed point.
Candidates can obtain additional marks by referring to a named developing country or countries they may have studied. For full marks both people and the environment should be mentioned.
Answers may include:

Some countries are too dependent on just one or two exports (1). Ghana, for example, depends on cocoa for 85% of its exports (1). If the price of cocoa were to fall, cocoa farmers would receive less income (1). They would have less to spend, affecting other businesses in Ghana (1). If the price of the export in a developing country was to increase, then there would be more income for the country to invest in other products and industries (1). Countries would be less affected if they had a greater range of products to export (1). This could lead to improvements in infrastructure, education and health (2).
Or any other valid point.   6

13 (a) 1 mark for a valid point.
2 marks for a developed point.
Answers must refer to differences. It is not necessary to refer to all of the parks for full marks.
Answers may include:

The most popular national park is the Lake District with 15·8 million visitors a year compared with 4 million visitors in Loch Lomond (1). There is also a large difference in the income for these two parks, with Loch Lomond earning £150 million as opposed to the Lake District's £952 million (1). The Cairngorms earn £185 million from their 1·5 million visitors, whereas the other mountainous area, Snowdonia, obtains a much larger income of £396 million from a total 4·7 million visitors (1).
Or any other valid point.   4

(b) 1 mark for a valid point.
2 marks for a developed point.
If no area is named, mark out of 4.
Answers may include:

In coastal areas such as Dorset, the government has tried to encourage more tourists to visit inland areas in order to reduce pressure on coastal areas (1). There are laws forbidding the dropping of litter, and fines are used to enforce them (1). Under EU regulations, clean beaches can be awarded a

Blue Flag (1). A number of scenic areas in Britain have been designated as National Parks, such as the Lake District and Loch Lomond (1). Future developments such as tourist hot spots can be monitored and controlled (1). Areas within the parks can be designated for car parks and other buildings (1). In mountain areas, money can be spent on improving footpaths which have been eroded over time by hillwalkers (1). Visitors to these areas can be educated to respect the natural environment through information centres (1). Sports activities, such as mountain biking and quad biking, can be restricted to certain areas (1). There are organisations, such as the National Trust, which make great efforts to protect the environment, culture and historic buildings in popular tourist areas (1).
Or any other valid points.   6

14 (a) 1 mark for a valid point.
2 marks for a developed point.
Answers may include:
Cholera outbreaks have been reported throughout the world but particularly in developing countries (1). Apart from Europe, outbreaks have been reported in central and southern Africa, India and islands in South-East Asia, islands in the Caribbean and parts of Oceania (2). In places such as the USA, Britain, Germany and Australia, there have been cases where people travelling from other countries have imported some cases of cholera (1).
Or any other valid point.   4

(b) 1 mark for a valid point.
2 marks for a developed point.
Answers must refer to both causes and prevention methods for full marks. Otherwise, award a maximum of 4 marks.
For malaria, answers may include:
Malaria is spread by the female anopheles mosquito (1). The disease is carried by the mosquitoes, which have been infected by taking blood meals from infected persons and then passing it on in their saliva when biting another person (2). The mosquitoes breed in areas of still water such as swamps or even in water barrels (1).
Methods to prevent/reduce the disease include the use of insecticides to kill mosquitoes (1); using anti-malarial drugs such as chloroquine to treat blood parasites (2); releasing water from dams to drown immature larvae (1); draining breeding grounds such as swamps (1); using small fish to eat larvae (1); using mustard seeds to drag larvae below the surface to drown them (1); using nets to protect people from being bitten by mosquitoes when sleeping (1).
Or any other valid point.   6

NATIONAL 5 GEOGRAPHY MODEL PAPER 3

Section 1: Physical Environments

1 (a) Arête – 472693
Corrie – 467677
U-shaped valley – 525594   3

(b) Full marks can be awarded for diagrams which clearly illustrate/explain the formation of a corrie.
Answers may include:
Snow fills a hollow on the side of a mountain and is compressed and turns to ice (1). Ice moves downhill under gravity (1). The corrie floor is abraded (1). The ice plucks rock from the back wall (1). The hollow is deepened by erosion (1). A lip is left as the ice loses power (1). Freeze/thaw action helps to steepen the back wall (1).
Or any other valid point.   4

2 (a) V-shaped valley – 473657
Meander – 447585
Tributary – 528595   3

(b) 1 mark for a valid point.
2 marks for a developed point.
A well-annotated diagram could obtain full marks.
Answers may include:
A V-shaped valley is the upper part of its course which is eroded by the river (1). It is created by down-cutting/vertical erosion (1) by corrosion and hydraulic action (1). The exposed sides are weathered, for example, by freeze/thaw action (1). Particles of rock are moved down the slope by the movement of rainwater and gravity (1) and also by being transported away by the fast-flowing stream (1).
Or any other valid point.   4

3 1 mark for a single point.
2 marks for a developed point.
For full marks, reference must be made to at least two of the land uses.
Maximum 1 mark for a grid reference.
Answers could include:
• Deer stalking – On the highest land where it is too cold and the soils are too thin even for trees to grow (2). Deer are nimble and so can cope with steep slopes and can survive on rough grazing (2).
• Forestry – Most of the land is above 250m where it is too cold for crops to grow (1) and the growing season is too short (1). Soils are acidic and rainfall is high, but coniferous trees can grow in these conditions (2). Many of the slopes are too steep to use machinery (1).
• Mixed farming/settlement – Land is lower and the climate is warmer, so is more suited to settlement and cultivation, producing a pattern of scattered farm houses (2). Arable farming can take place on the flat alluvial soils of the Peffer flood plain (1). Livestock can be grazed on the steeper, sloping land (1).
• Ancient fort – A good defensive site on top of a hill (1) from which the advancing enemy could be observed (1). It would have been much easier to repel them from here than from down by the river (1).
Or any other valid point.   5

4 1 mark for a valid point.
2 marks for a developed point.
The answer must explain the differences in weather.
Answers may include:

There is greater cloud cover in Cape Wrath than in Banbury because a cold front has just passed Cape Wrath whereas Banbury is in the warm sector (1). Wind speeds are higher in Cape Wrath, as shown by the isobars being closer together than in Banbury (1). Temperature is lower in Cape Wrath because of the cold front, whereas Banbury is in the warm sector, which brings higher temperatures (1). The cold north-westerly winds have brought snow and sleet in Cape Wrath, whereas there is little or no rain in Banbury because of the warm sector (1).

Or any other valid point. 4

5 1 mark for a valid point.
2 marks for a developed point.
For glaciated uplands and water storage and supply, answers may include:

Glaciated uplands in Britain contain lochs and lakes (1). These are used to store water and to supply water to towns and cities (1). Local climate and underlying geology are important, and glaciated uplands are very suitable in this respect (1). These areas are located in areas which have high average rainfall, such as the north and west of Britain (2). The rocks in these areas are mainly impermeable rocks such as granite (1). This allows storage reservoirs to hold water (1).

Or any other valid point. 4

Section 2: Human Environments

6 1 mark for a valid point.
2 marks for a developed point.
No marks for grid reference.
Accept yes/no answers.
Answers may include:
Advantages:

The area is flat for building houses (1). There are pleasant views over the water (1). It has good road communications, allowing easy access to the area by car (1). There is also a railway station close by, so residents can travel by train (1). There is a hospital near the site in case of accidents (1). The town of Dingwall is within walking distance for provisions (1). There are leisure activities close to the location, e.g. leisure centre, museum and a castle (546601) (2).

Disadvantages:

The housing will be close to a rifle range, which could be noisy and dangerous (1). There are works at 560585 which would be an eyesore (1). It is close to the water, so could be dangerous for children (1). The railway line is close by, which could be noisy (1).

Accept any other valid point. 5

7 (a) 1 mark for a valid point.
2 marks for a developed point.
Answers may include:

The basic link between GDP per capita and the number of births per woman is that the higher the GDP, the lower the number of births per woman (1). The lower the GDP per capita, the higher of number of births per woman (1). For example, Austria has both the highest GDP per capita ($31000) and the lowest births per woman (1), whereas Guinea has the highest births per woman (six) but the lowest GDP per capita ($3000).

Or any other valid point. 3

(b) 1 mark for a valid point.
2 marks for a developed point.
The number of births per woman is low where GDP per capita is high because:

Women have access to family planning and contraception (1). Women are better educated to follow careers (1). This means that they marry later and delay having children until they are older (1). Therefore they have smaller families (1). Many women work instead of staying at home to look after children (1). Infant mortality rates are low, therefore there is less need to have so many children (1).

The number of births per woman is high where GDP per capita is low because:

There is a lack of birth control and family planning (1). Poverty and lack of health care means that many children die in infancy (1). So, parents have more children in the hope that a few will survive (1). In poor countries, children provide a workforce and earn money for their families (1). A lack of pensions and social benefits means that children are needed to look after their parents in old age (1). Religious and social pressures encourage people to have more children (1). Women marry young and have larger families (1).

Or any other valid point. 6

8 1 mark for a valid point.

2 marks for a developed point.

Developed countries:

As there are fewer children, this affects the numbers attending school, leading to school closures (1). There will be a smaller number of people available for work (1). There will be more elderly people dependent on a smaller number of people to support them (1). More money will be needed to provide medical and social care for an ageing population (1).

Developing countries:

Large families may be forced to live in poverty (1). Less food may be available for the population (1). Less money will be available for medical care and education (1). More children may die in infancy (1). Literacy rates will be much lower than in developed countries (1). Less industry means that unemployment will be high, with large families suffering the effects most (1). Standards of living will be low for many families (1). However, large populations can produce a large workforce.

Or any other valid point. **6**

Section 3: Global Issues

9 (a) Both carbon dioxide and global temperature must be mentioned for full marks. Otherwise, mark out of 3.

Carbon-dioxide concentrations:

Carbon-dioxide concentrations rose from 300 parts per million in 1900 to almost 400 parts per million in 2010 (1). From 1900 to 1950 there was hardly any change, but since 1950 carbon-dioxide concentrations have steadily increased (1).

Temperature changes:

Temperatures continued to increase during the period 1900 to 2010 (1). During this period, global temperatures increased by 1·5 degrees (1). Global temperatures fell in about 1915 and in 1950 by 0·25 and 1 degree respectively (1).

Or any other valid point. 4

(b) 1 mark for a valid point.

2 marks for a developed point.

At least two ways must be mentioned for full marks. Otherwise, mark out of 5.

Answers may include:

Introducing laws to reduce the burning of forests (1); introducing replanting schemes where forests have been destroyed (1); reducing the use of sprays which include CFCs and ensuring that no CFCs are allowed to escape from fridges etc. on waste-dumping sites (1); reducing exhaust emissions containing lead and carbon dioxide by adding filter systems to vehicle exhaust systems and producing cars/lorries which use lead-free fuel (1); reducing the use of fossil fuels such as coal, oil and natural gases by introducing 'green' fuels such as HEP, wind power, solar power and other renewable energy sources (2).

Or any other valid point. 6

10 (a) 1 mark for a valid point.

2 marks for a developed point.

Answers may include:

The highest percentage increase in forest cover during the period 1985 to 2010 was in Vietnam with +38%, with the lowest percentage increase occurring in India with +6% (2). The second highest increase was in China, with +25% (1). The highest percentage loss was in the Philippines with −32%, and the second lowest percentage loss occurred in Malaysia and Laos, each with 7% (2). Other countries suffering a loss included Myanmar −18%, Indonesia −25%, Cambodia −20% and Thailand −10% (2).

Or any other valid point. 4

(b) 1 mark for a valid point.

2 marks for a developed point.

If no area is mentioned, mark out of 5.

Advantages:

In Brazil, large areas have been cleared by timber companies, and the hardwood has been exported abroad (1). This earns income for the country (1). Forests have been cleared to make room for new farmland (1) and settlements for the expanding population (1) and to increase food production (1). Forest is also destroyed for mineral extractions which are also sold to other countries (1).

Disadvantages:
  The habitats of wildlife are destroyed (1). Burning trees releases vast quantities of carbon dioxide into the atmosphere and may contribute to global warming (1). The homes of indigenous tribes are destroyed, as are their traditional culture and way of life (2). Plants which may contain cures for diseases are also destroyed (2). Poor farmers lose their land and may be forced to migrate to towns and cities to find employment (1).
  Or any other valid point.   6

11 (a) 1 mark for a valid point.
    2 marks for a developed point.
    Answers may include:
    Earthquakes occur along or near the boundaries of large plates which make up the earth's crust (2). These areas are known as plate margins (1). These are found in southern Europe, through the Middle East and into eastern and South-East Asia (2). There are also earthquakes stretching from Alaska down through the west coast of the USA, through Mexico and down the west coast of South America (2).
    Or any other valid point.   4

  (b) 1 mark for a valid point.
    2 marks for a developed point.
    If no area is named, mark out of 5.
    Answers may include:
    In the area of eastern Turkey in October 2011, immediate aid was required to help hundreds of trapped people (1). Emergency rescue teams were brought in from all over the world, using sniffer dogs and infra-red cameras to locate people trapped in collapsed buildings (1). Tents and blankets were given to people made homeless (1). Bottled water was given to people, as water supplies had been cut off (1). Medical help was given to injured people (1). Money was given by other countries to help rebuild schools, hospitals and clinics (1). Foreign governments and the UN gave aid to help rebuild the infrastructure – roads, water supplies, electricity supplies and bridges (1). Farmers were supplied with seeds and machinery to help grow crops for the future (1).
    Or any other valid point.   6

12 (a) 1 mark for a valid point.
    2 marks for a developed point.
    Answers may include:
    The country's main export consists of machinery and electrical goods (64%), and its main import is also machinery and electrical goods (32%) (2). It imports and exports the same percentage of manufactured goods, with 11% respectively (1). Similarly, the amount of exports and imports of chemicals is almost the same at 6% and 7% respectively (1). The only raw materials imported include oil (18%) and timber (3%) (1). It does not export any raw materials (1). The second highest export consists of vehicles at 16% (1).
    Or any other valid point.   4

  (b) 1 mark for a valid point.
    2 marks for a developed point.
    Answers may include:
    Developing countries are often overdependent on raw materials for export (1). Many are overdependent on just one or two raw materials. For example, Zambia is dependent on copper – 87% of its exports – and Ghana depends on cocoa for 80% of its exports (2). If the prices of these commodities fall, then the countries will receive much less income (1). This means less money to buy imports or to pay for public services such as schools and hospitals (2). Workers will receive less money in their pay (1). Standards of living will fall (1). The country may have to borrow money from richer countries (1). The country's debt will rise, and its economy may fail (1).
    Or any other valid point.   6

13 (a) 1 mark for a valid point.
    2 marks for a developed point.
    Answers may include:
    Apart from France and the United Kingdom, the general trend is that of an increase in arrivals during the period 2008 to 2010 (1). The number of arrivals for Malaysia and Mexico has remained almost the same during this period (1). The largest increases took place in Spain, the United States, China and Italy (1). The combined total of arrivals in European countries vastly exceeds the number of arrivals in the USA and China (1).
    Or any other valid point.   4

  (b) 1 mark per valid point.
    2 marks for a developed point.
    If no named country, mark out of 5.
    Answers may include:
    Eco-tourism, in countries such as South Africa, helps people by bringing more money into the economy (1). Eco-tourism helps to create employment in a range of activities including hotels, farming, transport and retailing (2). Eco-tourism helps to protect the culture and traditions of native populations (1). Money from tourism helps to improve the local and natural environment by ensuring that beaches remain clean, historic buildings and sites are maintained, and wildlife and safari parks are set up in, for example, many African countries (2). Eco-tourism is aimed at encouraging awareness of the ecological damage that tourist development often entails (1). Eco-tourists are encouraged to visit certain areas to gain an understanding of the lifestyles and culture of local people and an appreciation of the natural environment and local ecosystems (2).
    Or any other valid point.   6

14 (a) 1 mark for a valid point.
    2 marks for a developed point.
    Answers may include:
    The number of cases in the provinces has dropped dramatically from about 63 000 cases in 2000 to about 6000 in 2010 (1). The largest decrease has occurred in KwaZulu Natal Province, with a drop of over 40 000 in the period shown (1). In 2010, there were no cases of malaria in KZN province (1). The number of cases in the other two provinces decreased from about 21 000 in 2000 to about 6000 in 2010 (1).
    Or any other valid point.   4

(b) 1 mark for a valid point.
   2 marks for a developed point.
   If Malaria chosen.
   Methods include:
   Draining areas with stagnant water, for example swamps, to destroy the breeding grounds of mosquitoes (2); using insecticides such as Malathion to kill mosquitoes (1); using nets to protect people from mosquito bites while they are sleeping (1); using drugs such as chloroquine to control the disease (1); setting up village health centres to provide information and education programmes (1); releasing water from dams to drown mosquito larvae (1); introducing small fish in paddy fields to eat the larvae (1); adding mustard seeds to water to drag larvae below the surface and drown them (2).
Or any other valid point.   6

NATIONAL 5 GEOGRAPHY 2014

Section 1: Physical Environments

1. (a) u-shaped valley: 323143
      corrie: 326168
      arête: 309201

   (b) A diagram with fully explanatory labels may gain full marks.
      Snow collects in the north facing hollow of a mountain and the bottom layers turn to ice (1); the ice begins to move and the hollow is eroded (1); Rocks frozen on to the bottom of the ice scrape away at the base of the hollow (1) this is known as abrasion (1); ice plucking is when the glacier freezes on to loosened rock and pulls if free as the glacier moves (1); frost shattering may cause material to be incorporated into the ice (1); the ice melts leaving a tarn or corrie loch (1).
      Or any other valid point.

2. (a) Ox-bow lake: 427099
      Meander: 423107
      V-shaped valley: 297207

   (b) A diagram with fully explanatory labels may gain full marks.
      As water flows over a hard rock band on to a softer one it erodes the soft rock faster (1); this creates a step which the water has to flow down (1); the river creates a plunge pool at the bottom of this drop (1) where the water swirls around and undercuts the hard rock (1); this erosion could be due to hydraulic action (1) where the force of the water erodes the rock (1) or due to corrasion where the river's load causes the erosion (1); eventually the hard rock collapses because there is nothing left to support it (1) and the waterfall will move back (1); over time this creates a gorge (1).
      Or any other valid point.

3. One mark for first valid grid reference.
   If **forestry** chosen:
   There are suitable areas for forestry around Loch Ailsh (3110) (1) because the land is not too high, mostly under 300 metres (1) and there is access for vehicles from the A class road (1); there are a number of other tracks allowing lorries to take away timber (1) and the land is probably unsuitable for most other land uses as it is quite steep (1).
   Or any other valid point.

   If **water storage and supply** chosen:
   This is an excellent area for water supply as there are a number of natural lochs such as Loch Ailsh (1) and deep narrow valleys which could be dammed (1) such as the Abhainn Gleann na Muic at 393130 (1); there is not much evidence of human activity so there are less likely to be objections to a dam (1); there are many streams and rivers (such as in 2914) indicating plentiful water supply (1).
   Or any other valid point.

   If **recreation/tourism** chosen:
   This area would be good for recreation and tourists because there are lots of opportunities for hillwalking and climbing (1) such as on Ben More Assynt (1); there are chambered cairns eg 299103 (1) and other historic sites in the forests around Loch Ailsh (1); there are some roads for access such as the A class road/minor road (1)

but mostly there are no signs of human activity and this would be an added attraction for many visitors (1) as they would appreciate the magnificent and unspoiled scenery (1).
Or any other valid point.

4. No marks awarded for description of the weather.
   (a) A tropical continental air mass will bring hot dry weather in summer which could result in droughts (1); there might need to be hosepipe bans (1); grass might wither and die causing problems for livestock farmers (1); ice cream sales might rise (1) as people make the most of the sunny weather and head for the beach (1); it could be very hot and difficult to do physical work outside (1); heavy rain from thunderstorms might cause flash floods (1).
   Or any other valid point.

   (b) It is much windier at Ballycastle than Lerwick because the isobars are much closer together in Ireland (1); it is raining heavily at Ballycastle because it is next to the cold front whereas Lerwick is not close to any fronts (1); the wind direction is north west at Ballycastle and south east at Lerwick because the wind goes anticlockwise around the low pressure and the isobars show the approximate direction (1); the cloud cover is 8 oktas at Ballycastle because it is so close to the cold front where there is likely to be cumulonimbus cloud whereas Lerwick is not near any fronts so is less likely to have cloud (1); the temperature is warmer at Ballycastle as it is still in the warm sector whereas Lerwick is not (1); the weather is mostly worse in Ballycastle than Lerwick as it is closer to a depression (1). It is warmer in Ballycastle because the clouds trap the heat whereas it is colder in Lerwick because there are no clouds (1)
   Or any other valid point.

Section 2: Human Environments

5. For full marks answer must refer to a named city.
   If Glasgow chosen:
   Many old buildings like Candleriggs Fruit Market are no longer needed or suited to their original purpose so are converted into houses, hotels etc (1), to make use of the valuable space (1). These converted buildings also afford the opportunity for new land users to move in (1). Many areas like Buchanan Street have become pedestrianised to make it safer for shoppers (1) and reduce the number of vehicles in the centre as well as reducing pollution (2). Many small shops have gone out of business and have been replaced with larger chain stores (1) as these land users can afford the high rents of the city centre (1). Indoor shopping malls eg Buchanan Galleries are being built to try to encourage customers back into the town centre (1). They are not affected by the weather and easily accessible to more customers (1).
   Accept any other valid point.

6. Answers must refer to changes or trends.
   (a) The percentage of children age between 0 – 14 steadily decreases between 1982 to 2050 by 18% (1). The % of working age population between 15 – 59 increased between 1982 to 2000 by 9% (1) but is predicted to fall from 2000 to 2050 by 10% (1). The % of the population over 60 has risen constantly since 1982 by 19% (1) and the highest predicted increase of 9% between 2020 and 2050 (1).
   Accept any other valid point.

   (b) China used the one child policy to help reduce growth (1) the government took away benefits from families if they didn't follow this rule (1), eg increased access to education for all, plus childcare and healthcare (2). China encouraged the use of birth control methods like IUD's and sterilisation (1). Abortion is legal in China and is widely used (1). In recent years China has provided more education on birth control (1). Some countries like Indonesia introduced more free contraception (1). Some countries have used sex education (1). Some countries use tax incentives to encourage smaller families (1).
   Or any other valid point.

7. Answer must refer to the differences between the cities. The population of Mumbai starts off far higher than Glasgow in 1981 (1) a difference of 7.5 m (1). Mumbai's population is rising between 1981 and 2011 whilst Glasgow's population is dropping (1). The fall in Glasgow's population levels out in 2011 whilst the rise of population of Mumbai is at its steepest (1). Mumbai's population has reached 13.6m by 2018 whilst Glasgow is at its lowest at just under 0.7m (1).
   Or any other valid point.

8. Pesticides reduce disease producing better crops (1) and a surplus to trade (1). Fertilisers increase crop yields (1) this leads to better profits for some farmers (1) which can lead to an increase in their standard of living (1). Mechanisation means less strenuous work for the farmer (1) and is quicker and more efficient (1).
   GM crops produce a greater yield and are disease resistant so make a greater profit for the farmer (1) they can reduce the cost to the farmer of applying pesticides (1) and reduce the risk to his health (1). The growing demand for biofuels means higher crop prices and can result in the farmer getting a higher income (1) and create employment (1).
   Or any other valid point.

Section 3: Global Issues

9. (a) There is a flooding risk in South Asia (1). There is a risk of drought in Western Australia (1).
   Coastal areas and low lying land are at greatest risk of flooding (1), eg Bangladesh/Indonesia (1).
   Areas which lack water and are at risk of suffering drought are desert and semi-arid regions (1), eg Namib-Kalahari (1).
   Crop yields decrease likely in land on edge of deserts (1), eg Sahara/Sahel regions (1).
   Any other valid point/location.

   (b) Maximum of 4 marks for either human or physical causes.

   Physical causes:
   Variations in solar energy may affect global temperature (1).
   Variations in the earth's orbit around the sun may cause temperature changes (1). Sulphur dioxide gas & dust particles released in a volcanic eruption can affect amount of solar energy reaching the earth (1). Changes in ocean currents can effect temperature in different parts of the world (1).

Human causes:
The biggest contributor is gas released into the atmosphere from cars and burning fossil fuels (1) Cow dung and, decaying landfill produced harmful gases such as methane (1) which contribute to global warming (1). Deforestation causes CO2 level to rise because there are less trees to absorb it (1) and also burning trees increases the amount of CO2 in the atmosphere (1). Cooling units - CFCs and HCFCs previously used as coolants in fridges, freezers and air conditioners are another cause of climate change (1).
Any other valid points.

10. (a) Answer must refer to both temperature and rainfall, otherwise maximum 3 marks.
It's very cold in the Tundra and there is not much rain (1). Tundra regions have a cold desert climate – less than 250 mm rainfall/year (1). Tundra regions tend to get a small amount of rainfall each month (1). Highest temperature is 6 ° C in July (1) and lowest is minus 28 ° C in Jan/Feb (1).
Range of temperature in the Tundra is 34 ° C (1).
Any other valid point.

(b) Answer must refer to advantages and disadvantages. Maximum of 4 marks for either advantages or disadvantages.

eg Alaska
Advantages: The Arctic's undeveloped oil/gas/mining/forestry/fishery and other wildlife resources have the potential to provide enormous opportunity and wealth for the future (1). Oil has brought many benefits to the native people including jobs (1). Development raises their standard of living (1), improving education and medical facilities (1). Development of resources can help economic expansion (1).

Disadvantages: Any oil spill has serious impacts on habitat of seabirds, fish, and marine mammals (1) which could take decades to recover (1).
WWF concerned since there is no effective method for containing and cleaning up any oil spill in ice conditions (1).
The Gwich'in Indians are concerned about the threat to their culture and way of life (1). Polar bears are also threatened with extinction as they are forced out of their natural habitat (1). Developments such as oil pipe lines affect migration routes of caribou (1).
Any other valid point.

11. (a) Most volcanic activity is found on plate boundaries (1). Many volcanoes are located on the west coast of South America (1) where the Nazca plate meets the South American plate and is forced underneath it (1). Many volcanoes are found along the mountain ranges of the Andes in South America and the Rockies in North America (2).
Accept any other valid point.

(b) For full marks both the people and the landscape must be mentioned.
If **Japanese Earthquake** chosen:
The earthquake caused a tsunami which flooded the land (1). Cars, ships and buildings were swept away by the wall of water (1). Nuclear reactor plant damaged (1). Thousands of people living near the Fukushima nuclear power plant had to evacuate (1). A 10m wave struck Sendai, deluging farmland (1) and sweeping cars across the airport's runway (1). Fires broke out in the centre of the city (1). A ship carrying 100 people was swept away off the coast (1). A dam burst in north-eastern Fukushima prefecture, sweeping away homes (1). About four million homes in and around Tokyo suffered power cuts (1). Thousands of people killed (1) or made homeless due to buildings collapsing (1).
Or any other valid point.

12. (a) Europe dominated World Trade Exports with around 43% in 2005. (1)
This had dropped to around 38% in 2010. (1) Europe still remains the largest exporter in 2010. (1)
Asia had the second largest regional share of World Trade with around 27% in 2005 (1), growing to around 31% in 2010. (1)
Africa's share is low, around 3%, (1) but has grown by about 1%. (1)
North America's share has dropped from just under 15% in 2005 to around 14% in 2010. (1)
Or Any other valid point.

(b) Farmers are paid a fair wage for their work (1) and safer working conditions are promoted. (1) The 'middle man' is removed, (1) meaning more money is paid to the local people. (1)
Money from fair trade can be used to improve services in local communities (1) such as schools and clinics (1) which improves standard of living (1).
Or any other valid point.

13. (a) The USA has six out of the ten most popular tourist attractions in the world including Niagra Falls and Disneyland (1). The most visited tourist attraction is Times Square in the USA with 35 million visitors per year (1). Washington D.C. is the second most popular tourist destination with 25 million visitors (1). Trafalgar Square is the most popular tourist area in Europe (1). Notre Dame and Disneyland in Paris are the most visited attractions in France with 12 million and 10.6 million visitors a year (2). Disneyland, Tokyo is the most visited attraction in Asia (1). Four out of the top ten most popular tourist destinations are Disneyland/Disneyworld parks located on 3 different continents (1).
Or any other valid point.

(b) Maximum of 4 marks for either effects on people or environment.
People (positive): Local people are employed to build tourist facilities eg hotels (1) and work in restaurants and souvenir shops (1). Employment opportunities allow locals to learn new skills (1) eg obtain a foreign language (1) and earn money to improve their standard of living (1). Services are improved and locals can benefit by using tourist facilities such as restaurants and water parks (1). Better employment opportunities increases the local Governments' revenue as wages are taxed (1) so they can invest in schools, healthcare and other social services (1). Locals can experience foreign languages and different cultures (1) and can also benefit from improvements in infrastructure eg roads and airports (1).

People (negative): Tourist-related jobs are usually seasonal therefore some people may not have an income for several months (1) eg at beach and ski resorts (1). Large numbers of tourists can increase noise pollution and upset the peace and quiet (1). Local people may not be able to afford tourist facilities as visitor prices are often higher than local rates (1). Tourists can conflict with local people due to different cultures and beliefs (1). There is additional sewage from visitors which increases the risk of diseases like typhoid and hepatitis (2).

Environment (positive): The appearance of some areas can be improved by modern tourist facilities (1). Some tourists are environmentally conscious and can have a positive impact on the landscape by donating money to local projects which help to protect local wildlife (1), eg nature reserves (1). Tourist beaches are cleaned up to ensure they are safe enough for people to use (1) through initiatives like Blue Flag (1). Seas become less polluted as more sewage treatment plants are built to improve water quality (1).

Environment (negative): Land is lost from traditional uses such as farming and replaced by tourist developments (1). Traditional landscapes/villages are spoiled by large tourist complexes (1). Air travel increases carbon dioxide emissions and contributes to global warming (1). Traffic congestion on local roads increases air and noise pollution (1). Tourist facilities such as large high-rise hotels and waterparks spoil the look of the natural environment (1). Litter causes visual pollution (1). Increased sewage from tourists can cause water pollution (1). Polluted water damages aquatic life and their habitats (1).
Or any other valid point.

14. (a) Maximum of 2 marks if no reference to figures. More children under the age of 5 die in developing countries (1). African countries have the most number of child deaths under the age of 5 (1). Many African countries have between 100-199 child deaths under the age of 5 (1), eg Sudan (1). There are mostly between 10-49 child deaths per 1,000 in Asia (1), eg Russia and China (1). There are typically between 10-49 child deaths under the age of 5 in South America (1) except Bolivia (1) which has between 50-99 (1). The continents of North America and Europe have the least number of child deaths (1). There are less than 10 child deaths under the age of 5 in many developed countries (1), eg UK (1).
Or any other valid point

(b) For full marks reference must be made to developed and developing countries.
Health Education programmes have been introduced to limit the spread of AIDS in Developing and Developed countries (1). ARV drugs are also more freely available (1). Condoms are available for free (1) and TV and Radio advertising has been used to get the message across (1). Agencies such as the World Bank have made funding available to Developing Countries to tackle the disease (1). In Developed Countries, needle exchanges (1) and drug therapy programmes (1) have been introduced.

NATIONAL 5 GEOGRAPHY 2015

Section 1: Physical Environments

1. (a) Headland - 766356
   cliff - 690382
   bay - 674398

   (b) **Stack**
   Waves attack a line of weakness in the headland (1). Types of erosion include hydraulic action, corrosion and corrasion (1). Continuous erosion will open up the crack and it will develop into a sea cave (1). Further erosion of the cave, often on opposite sides of the headland, will form an arch (1). The roof of the arch is attacked by the waves until it eventually collapses (1). This leaves behind a free standing piece of rock called a stack which is separate from the headland (1).
   Or any other valid point.

   Bay
   Bays are formed due to differential erosion (1) where rocks along the coastline are formed in alternating bands of different rock types (1) eg sandstone and clay (1) and which meet the coast at right angles (1). Clay is a softer rock than sandstone so it is eroded more quickly (1). The waves erode the softer rock through hydraulic action, corrasion and corrosion (1) to form sheltered bays (1) which may have beaches (1). The harder sandstone areas are more resistant to erosion and jut out into the sea to form exposed headlands (1).
   Or any other valid point.

2. (a) Levée - 684466
   Meander - 708473
   V-Shaped Valley – 713410

   (b) **Meander**
   In the middle/lower course, a river flows downhill causing lateral erosion (1). The river contains areas of deep water and areas of shallow water, this results in areas of slower and faster water movement and this causes the current to swing from side to side (2). The river flows faster on the outer bank and erodes it (1). This forms a river cliff (1). The river flows more slowly on the inner bank and deposits some of its load (1). This forms a river beach/slip-off slope (1). Continuous erosion on the outer bank and deposition on the inner bank forms a meander in the river (1).
   Or any other valid point.

   V-shaped valley
   In the upper course, a river flows downhill eroding the landscape vertically (1). The river erodes a deep notch into the landscape using hydraulic action, corrasion and corrosion (1). As the river erodes downwards the sides of the valley are exposed to freeze-thaw weathering which loosens the rocks and steepens the valley sides (2). The rocks which have fallen into the river aid the process of corrasion which leads to further erosion (1). The river transports the rocks downstream and the channel becomes wider and deeper creating a V-shaped valley between interlocking spurs (2).
   Or any other valid point.

3. Answers will vary depending upon the land uses chosen.

For farming: Reads Farm(1) (at grid reference 728489) is an example of a hill sheep farm as the land is steep (1). As the land is higher up, the weather will be harsh and sheep can survive these conditions, especially in winter (1). The land is too steep for farm machinery to operate (1). The soil will be too thin for crops to be grown (1).

For tourism and recreation: The South West Coastal Path follows the top of the cliffs and allows tourists to enjoy a view of the coastal scenery (1) eg 727367 (1). There is a nature reserve for people who want to observe wildlife at 747405 (1). There is a golf course for golf enthusiasts at 668428 (1). There are various camp/caravan sites for people to stay whilst visiting the various attractions in the area (1).

4. Answers will vary depending upon the land uses chosen.

Problems between tourists and farmers:
In the Cairngorms, tourists can disrupt farming activities as walkers leave gates open, allowing animals to escape (1). Tourists' dogs can worry sheep if let off their lead (1). Stone walls are damaged by people climbing over them instead of using gates/stiles (1). Noisy tourists can disturb sheep especially during breeding season (1). Farmers may restrict walkers access at certain times eg lambing season (1). Farm vehicles can slow up tourist traffic on roads (1) and parked cars on narrow country roads can restrict the movement of large farm vehicles (1).

Problems between industry and tourists:
Tourists want to see the beautiful and unusual scenery of the Yorkshire Dales but quarries spoil the natural beauty of the landscape (1). Lorries used to remove the stone endanger wildlife and put visitors off returning to the area (1). This threatens local tourist-related jobs eg in local restaurants (1). The large lorries needed to remove the quarried stone cause air pollution which spoils the atmosphere for tourists (1). Lorries cause traffic congestion on narrow country roads which slows traffic and delays drivers (1). The peace and quiet for visitors is disturbed by the blasting of rock (1). Some wildlife habitats may also be disturbed by the removal of rock (1).
Or any other valid point.

5. South-East England is usually warmer because it is closer to the Equator (1). This is due to intense heating from the sun (1) because sun rays are more concentrated (1). Places in Northern Scotland eg Wick, are colder because they are closer to the North Pole(1). This is due to a lack of insolation from the sun as the rays are less concentrated (1) and reflection of heat by the snow and ice (1). Places located on flat low-lying land are warmer eg Central Scotland, because temperatures increase as altitude decreases and places higher up ie mountainous regions are colder (1) because temperature decreases by 1oC for every one hundred metres in height (1). Places which are south facing are warmer because they get more sun (1) and places which are north facing are colder because they experience cold northerly winds (1). Western coastal areas are warmer because of a warm ocean current (1) (The North Atlantic Drift) and due to the prevailing South-Westerly winds that are warmed as they pass across this warm ocean current (1). In summer, places closer to the sea are cooler and in winter they are warmer because the sea heats up slowly in summer and cools slowly in winter (2).
Or any other valid point.

Section 2: Human Environments

6. (a) Main roads lead into this square (1) there is a bus station (1) and two railway stations (1) tourist information centre (1) several churches (1) museum (1) Or any other valid point

 (b) The land is flat so easy to build on (1) there is space available for expansion (1) eg expansion of the motor works at 163823 (1). There are good transports links like the M42 allowing people and products access to and from the area (1). A rail link with Birmingham International Rail Station gives easy access to the airport (1). There are many road junctions and intersections connecting the area to other areas and less traffic congestion as it is away from Birmingham city centre (2). The land is on the edge of Birmingham so will be cheaper encouraging housing estates like Sheldon to be built (1). The cheaper land allows the houses to be bigger with cul-de-sacs, gardens etc.(1) the houses can provide a source of labour for the airport, motor works and the business park (1)
 Or any other valid point

7. Contraception and family planning is widely available (1). Later marriages are more common which results in fewer children (1). People no longer choose to have lots of children as improved medical care and advances in medicine (1) has resulted in most children surviving at birth (1). Developed countries have the money to invest in medical care which reduces the infant mortality rate thus causing the birth rate to fall (1). Children are expensive so the greater number of children the bigger the financial burden (1). Women want careers so put off having children to a later age (1) or limit the size of their families to give them a reasonable standard of living (1). Sex education in schools helps to lower birth rates (1).
Or any other valid point.

8. For example in Rocinha (Rio), the former wooden shacks have been upgraded to permanent dwellings with some modern services (1). Residents constantly improve their homes through a process of 'self-help' (1) where the residents are provided with materials like bricks (1). Some prefabricated houses have been built by the Brazilian government (1) with basic facilities like toilets, electricity and running water (1). The residents have been given the legal rights to the land (1), roads have been built into/ or improved in the favela (1) allowing services like rubbish collections to take place (1), there are now a few health clinics and schools provided (1).
Or any other valid point.

Section 3: Global Issues

9. (a) The overall trend is that the amount of Arctic Sea ice has decreased between 1979 and 2013 (1) from (around) 7 million square kilometres to (about) 5 million square kilometres (1). There has been a fluctuation in the extent of sea ice in certain years (1) eg in 2013, the amount of sea ice increased from 3.75 million square kilometres in 2012 to 5 million square kilometres (1) whereas, between 2006 and 2007 there was a sharp decrease (1) from 6 million square kilometres to 4.25 million square kilometres (1).
Or any other valid point.

(b) Increased temperatures are causing ice caps to melt so Polar habitats are beginning to disappear (1). Melting ice causes sea levels to rise (1) threatening coastal settlements (1). An increase in sea temperatures causes the water to expand, compounding the problem of flooding (1). Global warming could also affect weather patterns, leading to more droughts (1) crop failures and problems with food supply (1); flooding, causing the extinction of species (1) and more extreme weather, eg tropical storms (1). Tourism problems will increase as there will be less snow in some mountain resorts (1). Global warming could threaten the development of developing countries as restrictions on fossil fuel use may be imposed to slow the rate of increasing CO_2 levels (1). In the UK, tropical diseases like malaria may spread as temperatures rise (1). Plants growth will be affected and some species will thrive in previously unsuitable areas (1). Higher temperatures may cause water shortages (1).
Or any other valid point.

10. (a) Overall the amount of deforestation in Peru 2004-2012 has decreased (1) from just under 3 million ha to 750 000 ha (1). The deforestation rate declined rapidly from 2004 to 2007 (1). Deforestation increased from 2007 to 2008 peaking in Peru at 1 500 000 hectares per year (1). Again Peru experienced a decline in deforestation rates from 2008 to 2009 by over 500 000 ha (1). From 2009 to 2010 deforestation rates rose to around 1 400 000 ha (1), before declining to around 750 000 hectares per year in 2012 (1).

(b) New industries have led to the expansion of towns such as Anchorage in Alaska which have grown to accommodate workers (1). Although these industries provide employment (1), these developments spoil the appearance of the natural landscape (1). New roads have been built to transport people and goods. This increases the number of vehicles in the tundra creating noise and air pollution (1). But also improves access to locals (1).

Oil is a very important industry in Alaska. The building of oil platforms and oil pipelines has resulted in damage to tundra vegetation and wildlife (1). In some areas, the Trans-Alaskan oil pipeline has been built on natural migration or hunting routes for animals, which hinders the natural movement of caribou (1). Local Inuit people have also had their way of life disrupted as they must detour around the pipeline (1) and may no longer have access to their traditional hunting grounds (1).
Local people were promised jobs in the industry, but few jobs are available for locals (1).

Burst pipes have spilt hundreds of thousands of gallons of crude oil in Alaska, devastating this fragile environment (1). Oil spills have also been responsible for pollution in the region (1), such as the Exxon Valdez disaster (1).

Any damage to the tundra landscape is slow to recover, as the short growing season means that bulldozer tracks from the oil and natural gas industries could take centuries to restore (1).

Pollution from mining and oil drilling has contaminated the air, lakes and rivers (1).
Any other valid point.

11. (a) Most cities are located on or near plate boundaries (1) where seismic activity is highest (1). Most earthquake threatened cities are found in developing countries (1) like Indonesia (1). A large number of threatened cities are found in China (1). Three cities in Africa are at risk (1). All threatened cities in the USA are found on the west coast (1) with a cluster around San Francisco/Los Angeles (1).
Or any other valid point.

(b) In Japan people take part in earthquake drills to practise what to do in the event of an earthquake (1) giving them a better chance of survival (1). The government warn people, using text messages and TV, giving them the chance to move to a safer place (1). Earthquake resistant buildings reduce the number of people trapped or killed (1). Sprinkler systems and gas cut off valves prevent fires spreading reducing the number of people injured and buildings destroyed (1). People living in earthquake prone areas have emergency plans in place and emergency supplies such as bottled water and tinned food are stockpiled to ensure they have vital supplies to survive in the event of an earthquake (2). In the event of an earthquake short term aid in the form of food, medicine and shelter is sent to the area to treat the injured (1).
Or any other valid point.

12. (a) The value of exports from developed world countries to developing world countries is $738bn (1) whereas there is only $650bn worth of goods exported from developing to developed world countries (1). That is a difference of $88 billion (1). The value of trade between developing world countries is $383bn (1). The value of trade between developed world countries is $2251bn (1). There is more trade between developed world countries than between developing countries (1); it is $1868bn more (1).
Or any other valid point.

(b) There is a big imbalance in the pattern of trade between the developing and developed world; this can reinforce differences in wealth between areas such as the EU and Africa (1); African countries export mainly primary products such as oil or cocoa beans for comparatively low prices but import mainly processed goods such as vehicles for much higher prices (1) which can result in a trade deficit for them (1); this can increase levels of poverty within African countries and cause difficulties for the economy as well (2); often the producers such as cocoa farmers in Africa receive very low wages and so struggle to maintain a decent standard of living (2); wealthy European countries profit from selling expensive manufactured goods to African countries (1), helping to keep a much higher standard

of living for their citizens (1); often, exploitation of primary products in African countries can lead to serious environmental damage, such as logging which has caused deforestation (1), resulting in the loss of areas of rainforest as well as the destruction of animal habitats (1).
Or any other valid point.

13. (a) There has been a fairly steady increase in visitor numbers since 1995 (1) from around 525 million reaching 1 billion in 2013 (1). There were only 2 years where the numbers decreased slightly ie in 2003 (1) when it dropped to just under 700 million (1) and in 2009, dropped to under 900 million (1). The period with the largest increase was the 5 years between 1995–2000 (1) whereas the slowest increase has been in recent years from 2010 (1).

(b) If Costa Rica cloud forest chosen:
Eco-tourism raises local as well as international awareness of natural environment (1) such as wildlife and vegetation (1). Developing countries now want to conserve their fragile environments and view eco-tourism as a significant means of generating income (1). Developed countries want to help developing countries conserve their fragile environments by promoting sustainable/eco-tourism (1). Tourists are now more environmentally conscious and want to help protect fragile environments for future generations (1). Eco-tourism provides work and opportunities for local people (1) hence improving their standard of living (1) encourages local enterprise and improvement schemes (1) promoting awareness of local culture and traditions (1).
Or any other valid point.

14. (a) Male deaths from heart disease are most common in Eastern Europe (1). Russia for example, has a rate of 444–841 per 100 000 (1). This compares to only 120–238 in the UK (1). Canada, the USA and Mexico have some of the lowest rates (1), with under 120–238 per 100 000 (1). Many central African countries have rates of 363–443. (1)
Or any other valid point.

14. (b) If **pneumonia** chosen:
Antibiotics are used to treat any bacterial lung infections (1) and patients are encouraged to drink plenty in order to avoid dehydration (1); in severe cases a drip may be required to restore the right level of salts and fluids quickly (1); paracetamol is used to ease the effects of fever and/or headaches (1); introducing more community-based health workers helps to control the incidence of pneumonia as children with the disease are more likely to be diagnosed and treated quickly (1); this can often help to save lives (1). Vaccinations are being increasingly used in developing world countries to protect children against common infections such as flu (1); adequate nutrition helps to increase a child's natural defences against disease and so education about this also helps to reduce pneumonia (1).
Or any other valid point.

If **kwashiorkor** is chosen:
The main method of managing kwashiorkor is education about the need for a well-balanced diet, so that children don't develop the disease in the first place (1); by educating communities they can be encouraged to grow different food types to increase protein intake (1); this might include crops such as cashews, peanuts, lentils or sunflower (1) and might also involve advice about constructing irrigation schemes to help crops grow better in times of drought (1); education about family planning also helps to reduce the number of children per family, making more food available per child (1).

For children who have kwashiorkor it is important to give vitamin and mineral supplements as salt and mineral levels in their blood stream may be dangerously low (1); Zinc supplements might also be administered to help the skin recover (1). Small amounts of food are reintroduced slowly, such as carbohydrates to give energy (1) and protein rich foods to help the child's body recover (1).
Or any other valid point.

If **malaria** chosen:
Anti-malarial drugs kill blood parasites (1) Chloroquine is an example of this. (1) Insecticides, such as malathion destroy the female anopheles mosquito (1).

Draining all breeding areas eradicates larvae (1), planting eucalyptus trees to soak up moisture removes breeding ground (1). Water can also be released from dams to drown immature larvae (1). Mustard seeds can be used to drag larvae below the surface to drown them (1). Small fish can be introduced to eat larvae and provide a cheap protein source (1). Genetic engineering of sterile male mosquitoes reduces mosquitoes (1).

Health education teaches people about how to protect themselves from being bitten (1). Preventative bed nets are cheap and effective at stopping mosquitos biting at night (1). New treatments have also been developed which seem to be more effective such as artemesinin/ACT because malaria parasite is not yet resistant to them (1).
Or any other valid point.

If **cholera** chosen:
One of the main ways to reduce or control the spread of cholera is to improve sanitation which stops disease from spreading (1). Providing wells and pipes makes drinking water safe and clean (1). Health Education encourages people to wash hands often with soap and safe water preventing infection as does building and use of latrines (2). Because of contaminated water people should cook their food well and eat it hot (1). Food stuffs should be kept covered and fruit and vegetables should be peeled to prevent contamination (2).

Cholera is an easily treatable disease.
The main ways to treat cholera are either a simple drink made from 1 litre of safe water, 6-8 teaspoons of sugar and 1/2 teaspoon of salt helps to rehydrate sufferers so that they can fight off the disease (2) or re-hydration tablets, if available (1). In especially severe cases, intravenous administration of fluids may be required to save the patient's life (1). Treatment with antibiotics is recommended for severely ill patients to help fight the infection (1).
Or any other valid point.

Acknowledgements

Permission has been sought from all relevant copyright holders and Hodder Gibson is grateful for the use of the following:

Image © gary yim/Shutterstock.com (Model Paper 1 page 6);
The diagram 'Land Use in the Rainforest' © Rhett A. Butler/Mongabay.com (Model Paper 1 page 10);
The diagram 'Global Tourist Arrivals 1990–2011' adapted from: http://en.ria.ru/infographics/20120928/176277103.html?id= © RIA Novosti/UNWTO (United Nations World Tourism Organization) (Model Paper 1 page 13);
Image © Oliver Hoffmann/Shutterstock.com (Model Paper 1 page 14);
Image © Africa Studio/Shutterstock.com (Model Paper 1 page 14);
The diagram 'Average Global Temperatures 1880–2012'. Data source: NASA Goddard Institute for Space Studies. Image credit: NASA Earth Observatory, Robert Simmon (Model Paper 2 page 9);
The diagram 'Distribution of Tropical Storms' © Caitlin.m/Creative Commons (CC BY-SA 3.0) http://creativecommons.org/licenses/by-sa/3.0/deed.en (Model Paper 2 page 11);
The diagram 'Cholera outbreaks 2010–2011', taken from http://www.soziologie-etc.com/med/ziv-u-korr/medizinkartell/02-d/008-weltkarte-cholera-ausbrueche2010-2011.gif. Data Source: World Health Organization. Map Production: Public Health Information and Geographic Information Systems (GIS) World Health Organization © Geneva, World Health Organization, 2012 (Model Paper 2 page 14);
The diagram 'World Population Growth (projected)', taken from United Nations (U.N.) Population Division, Long-Range World Population Projections: Two Centuries of World Population Growth, 1950–2150 (U.N., New York, 1992), p 22 (Model Paper 3 page 8);
The diagram 'Distribution of Earthquakes, October 2012'. Image courtesy USGS (public domain) (Model Paper 3 page 12);
Image © Alan Scott (http://www.rampantscotland.com) & Gianni Versace S.p.A (2014 page 6);
Image © D.J. Alexander (2014 page 6);
Image © Ben Cooper (2014 page 6);
Image © Alis Leonte/Shutterstock.com (2014 page 13);
Image © Simon Rawles (2014 page 17);
Image © jan kranendonk/Shutterstock.com (2014 page 19);
Image © Gary Whitton/Shutterstock.com (2015 page 10);
Image © Kanokratnok/Shutterstock.com (2015 page 10);
Image © markobe/Fotolia (2015 page 11);
Image © Ivan Cholakov/Shutterstock.com (2015 page 12);
Image © Lledo/Shutterstock.com (2015 page 12);
Ordnance Survey maps © Crown Copyright 2015. Ordnance Survey 100047450.

Hodder Gibson would like to thank SQA for use of any past exam questions that may have been used in model papers, whether amended or in original form.

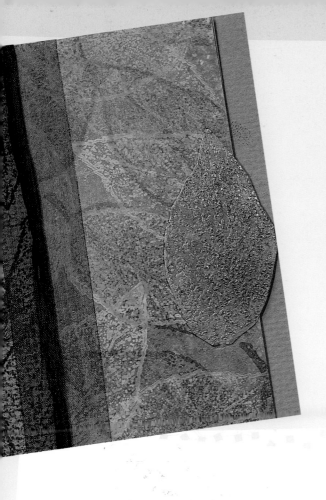

dedication

To all the customers I see

throughout the country

at classes and shows—thanks!

acknowledgments

Thanks to Shirley Hardee and Debra Valoff for their endless encouragement during this particular book. I owe a huge debt of gratitude to my editor, Krista Hamilton, for her tireless effort. Thanks again to the North Light crew for the pure pleasure of working with you guys!

contents

THE PROJECTS, 20

iNTRODUCTiON

there is no other communication so dear as a sincere greeting card. Whether it is for a holiday, a special event or even just a thoughtful note, a greeting card is like a virtual hug. And a handmade card is even more special! In this day and age, knowing you took the time to create something just for that special someone is heartwarming.

Creating your own greeting cards can be very rewarding. And since the price of an especially nice printed greeting has skyrocketed, it certainly makes sense to create your own. Many cards can be made simply with a few products and papers that can be used over and over. That said, my advice is to buy the best products you can afford. After all, handmade items reflect the person who has created them. Quality shows, and in the long run, it saves time, money and effort when used effectively.

In this book, I have included a variety of creative techniques. You'll find a three-dimensional box card perfect for holding small treats, a Thanksgiving card made with a hand-carved turkey stamp, a pop-up card with a surprise inside and many more. The projects require some common craft supplies like cardstock and craft scissors, and also some supplies that may be new to you, like pewter stickers and acrylic tiles. That's part of the fun! All you need is a little time, a lot of imagination and a quick trip to your favorite stamping supply store. Soon, you'll be on your way to creating greeting cards for every occasion. Enjoy!

GETTING STARTED

everyone, I think, starts with the belief that they are saving money when they begin to create their own cards. While this is definitely true in the long run, an initial investment must be made on tools and supplies. The real reason for making your own greeting cards is to fulfill your own need to create something wonderful for someone special. Knowing this makes the time and effort you put into a project more precious.

Before you begin, there are a few basic things you need to know for making your own greeting cards. I like to think of these simple techniques as the building blocks. Are they absolutely necessary? No. But doing things properly the first time will make the entire experience much more enjoyable.

Basic Tools & Materials

Tools

There are a few tools that always come in handy when I'm making greeting cards or working on another type of crafting project.

Bone folder

The bone folder is a great tool for scoring paper and smoothing down creases. Bookbinders use it for turning corners and scoring. Some bone folders are actually made from bone as the name suggests, while others are made from resin or wood. They come in a variety of lengths and are very helpful in all kinds of crafts.

Brayer

Brayers come in so many varieties, it's hard to choose which to buy. For my money, the best all-around brayer is the detachable 4" (10cm) soft rubber brayer. It will handle most jobs and is much easier to clean than other brayers. You will also find sponge, acrylic, hard rubber and wooden brayers. Each yields a different result. Try them all and see which one you like the best.

Craft knife

A craft knife is an invaluable tool when creating greeting cards and other stamp projects. X-acto is the most common brand. The blade should be very sharp and should be changed often to ensure clean cuts. You will learn more about the proper way to hold a craft knife on page 14.

Craft scissors

A good pair of stainless steel craft scissors are a must for most craft projects. They should be light-weight and have a comfortable grip for easy maneuverability when cutting out fine details.

Double-stick tape

Double-stick tape comes in a variety of forms and is available at art supply stores. In this book, I have used regular double-stick tape and dimensional double-stick dots.

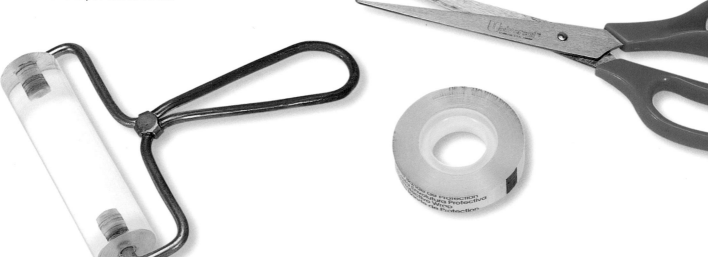

Embellishments

Beautifying your cards with decorative embellishments is the best part about card-making. Browse the aisles of your local craft store and experiment with all the different kinds of paint, markers, crayons, powders, pens and more.

Embossing powder

Embossing powder is required for many of the cards in this book. To use it, stamp an image with pigment or embossing ink. Dust the powder over the wet ink and shake off the excess. Use an embossing tool (heat gun) to melt the powder and create a raised design. Embossing powder comes in metallic, solid and multicolor forms.

Water-soluble crayons

These are available in stamp and art supply stores. I prefer soy-based crayons because they have a creamy texture and are loaded with pigment.

Powdered pigments

These are raw pigments used for a variety of purposes, including making your own paints. You can also use these pigments as a surface coating on paper or collage projects. Powdered pigments do need what is known as a "binder" to keep them adhered to your project. In this book, we will be using Diamond Glaze as a binder. Other options include white glue, paint media, gum arabic or spray fixative. Mix any of these with the powdered pigments to create a colored medium you can apply to many surfaces.

Thread, beads and cords

Decorative accessories such as thread, beads, paper cord, tassels and gift tags can be found at most stamp stores. I also find these items in specialty stores for beads and needlecrafts. Office supply stores are great for unusual items, too. Keep your eyes peeled because you never know what you'll find in the most unlikely places.

Paper

Always buy the best paper you can afford. When I lay out paper for a class, inevitably people choose the most expensive pieces first. Can you guess why? Great paper looks and feels fabulous. Paper is where I spend the bulk of my money every year.

Cardstock

I have used cardstock—a heavy, textured paper—for most of the greeting cards in this book. Cardstock comes in all sorts of colors and textures and is durable enough to withstand a little wear and tear.

Vellum

Vellum is a translucent paper with a smooth finish. It comes in a variety of weights, colors and patterns. Lighter weight vellum is easier to cut and score, but heavier vellum works better for heat embossing and painting.

Acetate

Although acetate is considered a plastic rather than a paper product, it can be used in a similar way as vellum. The clear plastic can be found in most stamp stores. Be sure to get embossable acetate (also known as "window plastic") in case you want to heat the piece. The thicker the laminate, the better it will work for the projects in this book because of the beating the pieces will take.

Stamps

There are millions of stamp designs out there, and again this is the time to invest in quality. I prefer well-trimmed, thick red rubber mounted on wood, or deeply etched unmounted rubber. Foam stamps are great for kids and temporary projects, but as a collector and an artist, I want a stamp that will hold up to use, abuse and time. All of the companies listed in the Resource Guide (pages 124 - 125) make quality stamps.

Ink

When it comes to ink, things can be very confusing. There are so many inks that perform a multitude of tasks, and they are available in every color under the sun! Let me break it down for you.

Dye ink

Most of the time, dye inkpads come with solid lids over the tops. This is because the ink is translucent and you cannot tell what the true color is until it has been stamped. Dye inks work on all types of paper and many porous surfaces. They are best for beginners and produce the most vibrant colors. Most dye inks will fade to some extent, and they tend to dry quickly, making them a poor choice for embossing. This also means dye inkpads can dry out quickly if you live in a dry climate.

Pigment ink

Slower-drying pigment ink is opaque and resembles paint. This ink is used for embossing or foiling with dry pigments. It is best suited for porous, uncoated paper. There are several brands, such as Brilliance, that dry on shiny and slick paper, but it takes a long time. To speed up the process, dry pigment inks with an embossing tool (heat gun). Since pigment inks are opaque, many are metallic or pearlized.

Solvent ink

Solvent ink, also called permanent ink, is made specifically for nonporous surfaces, but it also works well on paper. When dry, solvent ink does not smear—even when water is applied—making it perfect for watercolor techniques and markers. Since solvent ink works on every surface, it is my personal favorite.

Here is my basic advice. If you have never bought a single inkpad, buy a dye inkpad, a pigment inkpad and a solvent inkpad—each one in black. Then, buy your favorite metallic (gold, silver, copper or bronze) color in pigment ink and ten good colors of dye ink. These supplies should see you through most projects. After that, you are on your own in this addiction to color! I still buy almost every color of most of the brands out there.

Basic Techniques

There are a few basic techniques you'll need to know before you begin. These little tricks, which I've picked up over the years, will make every project easier and more enjoyable.

How to Hold a Craft Knife

The best way to make precise cuts with a craft knife is to control your blade. A sharp blade is also important, so be sure to replace it often.

1. Position the Knife
Hold the craft knife like a pencil and point the blade straight down toward the paper.

2. Make the Cut
Push the point of the blade into the surface, and then pull it down as far as you can, holding the blade parallel to the surface as you cut.

How to Hold a Ruler

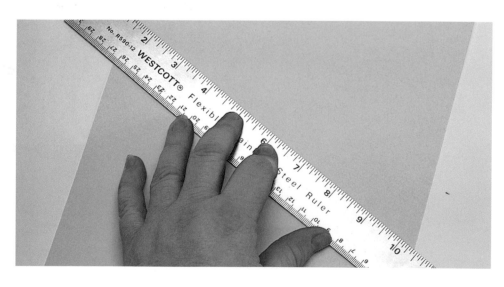

Holding your ruler correctly will make all the difference between a good card and a great card. Follow this technique and see for yourself!

Position the Ruler
Hold the ruler with your ring finger and thumb firmly positioned on the outside edge. This will prevent the ruler from slipping. Place your index and middle fingers directly on the ruler.

How to Ink a Large Stamp

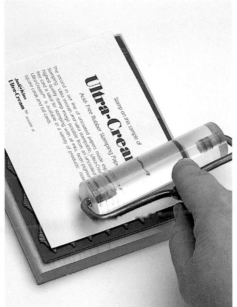

Nothing is more frustrating than lifting the stamp from your paper, only to find half the image didn't transfer. You must be sure to ink your stamp properly, especially when using large stamps like some of the ones in this book. Keep trying. In the stamping world, you get plenty of chances to make a good impression.

1. Ink Stamp
For the best coverage, rub the surface of the stamp with a pad of ink.

2. Transfer Image
Keeping the stamp facing up, position the cardstock on the stamp and roll a soft rubber brayer over the back of the cardstock to transfer the image.

How to Clean Your Stamps

When cleaning your stamps, use a cleaner with conditioner in it. I prefer a solvent-type cleaner, which removes any ink or paint I might be using. I personally do not like to overclean my stamps. This can damage the cushion under the die and loosen the adhesive.

1. Apply Cleaner
Apply a generous amount of stamp cleaner to the rubber stamp.

2. Blot Stamp
Blot the stamp on a clean, dry paper towel and reuse as necessary or store.

How to Score Paper

Scoring is especially important on heavy cardstock. Many tools can be used, such as a bone folder, paper clip, ballpoint pen, stylus or the back of a craft knife. Using your ruler as a guide, press down hard to get a nice, crisp crease.

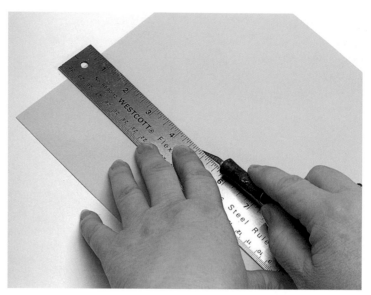

Score with Craft Knife
For a sharp crease, score with the back of a craft knife.

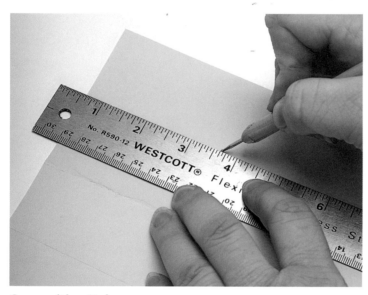

Score with a Stylus
For a soft crease, score with a stylus.

Always Use Tools
Never crease the paper by hand to score it.

TIP

For the most impressive pop-up and tri-fold cards, crisp scoring is a must!

How to Carve a Stamp

Carving your own designs into rubber is fun and addictive! Stamp carving tools are available in a variety of tip sizes, ranging from extremely thin to extremely wide. Soft white or pink rubber erasers work best for carving. You can even order large carving blocks up to 4' x 6' (123cm x 183cm) if you really get into it!

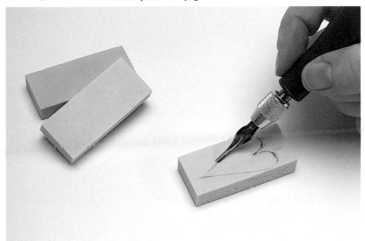

1. Draw and Trace Design
Draw your design in pencil on a soft rubber eraser. Trace the outline of the design with a thin carving tool.

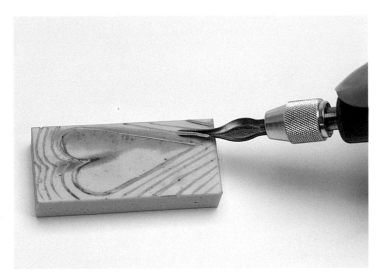

3. Carve Stamp
Carve around the stamped image with carving tools. Remove large chunks with a wide tip, and carve fine details with a thin tip.

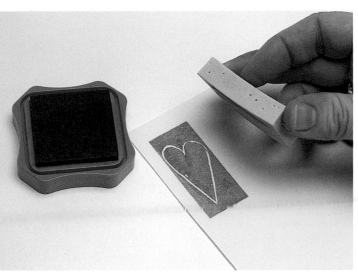

2. Ink Stamp
Ink the stamp and stamp it onto paper. This will indicate the areas that still need to be carved and cleaned up.

4. Ink Stamp
Ink the stamp and stamp it onto paper again.

How to Carve a Stamp (cont.)

5. Cut Away Excess
Cut away the unwanted portions of the eraser with the craft knife.

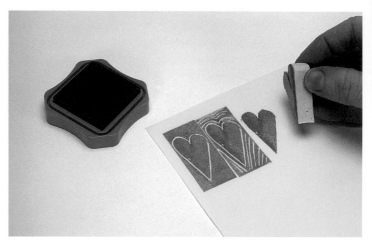

6. Stamp onto Paper
Stamp the inked image onto paper again to test it out.

How to Remove Smudges

We all make mistakes from time to time. It's part of the learning process. Accidental smudge marks can be easily fixed with a regular craft knife and a white eraser.

1. Scrape Paper
Using your craft knife, gently scrape the surface of the paper over the error. Be careful not to tear the paper.

2. Erase Rest
Rub the area with a white eraser to remove the rest of the smudge.

How to Make a Mask

Masking is an important technique to master, and practice makes perfect! In regular masking, the mask covers up the object, and the images are stamped around it. For an example of regular masking, see "In the Mail" on page 74. In mortise masking, the area around the object is masked, and the object itself is stamped. For an example of mortise masking, see "Comfy and Cozy" on page 112. Try both and see which technique you like best.

Many crafters use Post-it notes for masking because they are lightly adhesive and available in many sizes. I prefer Eclipse tape by JudiKins, and have used it for the projects in this book. This product is sold by the roll and is lightly adhesive on one side.

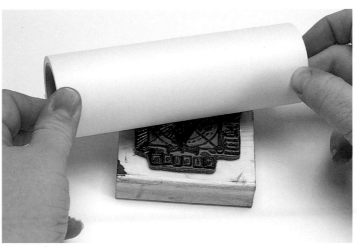

1. Stamp on Eclipse Tape
Ink a stamp and roll Eclipse tape over it.

2. Cut Out Eclipse Tape
Cut out the portion of the stamped Eclipse tape you want to mask off. For a regular mask, use the object itself. For a mortise mask, use the area around the object.

3. Stamp Paper
Stamp the paper with colored inkpads.

4. Ink and Stamp
Ink a stamp with colored ink and stamp over the colors on the paper. Remove the mask to reveal a stamped card.

THE PROJECTS

now that you know the basics of card-making, you're ready to begin. The cards in this book are broken into five categories: Winter, Spring, Summer, Fall and Timeless Occasions. As you follow my step-by-step instructions, feel free to use your own stamps, colors and materials, and try not to limit yourself by the season. By swapping out a few stamps and a piece of cardstock, the Christmas Box card can easily become a tiny box to hold wedding favors. I hope you use the techniques shown in each project as a springboard for more fun and creative ideas. Remember, there is no right or wrong way to make a handmade greeting card. The more personalized it is, the better!

CARDS FOR WINTER

Winter is the time of year when families and friends get together to celebrate holidays and special occasions. What a perfect opportunity to create greetings that will make them glow with delight. I can think of nothing better than curling up by the fire on a cold winter's night with only my card-making supplies and my creativity. A simple, handmade greeting card may be the small gift that warms the heart of someone you love this season.

CHRISTMAS BOX

this is a simple and fun card to make for many holidays. The three-dimensional box can be filled with candy, small toys, gift certificates or money, which makes it perfect for Christmas or Hanukkah! There are many template stamps like this one available at your local stamp store.

TIP

This is an easy card to replicate over and over if taken in steps. I make holiday cards in bulk, completing one step each night. That way, I stay focused and get several cards done at once!

What you'll Need

Stamps*

- Box template stamp
- Present stamp
- Christmas tree stamp

Materials

- 3 pieces of green cardstock
- White cardstock
- Clear acetate square
- Black dye ink
- Red dye ink
- Green dye ink
- Clear-drying glue
- Glitter
- Double-stick tape
- Craft scissors
- Craft knife
- Ruler

*See Resource Guide for stamp credits.

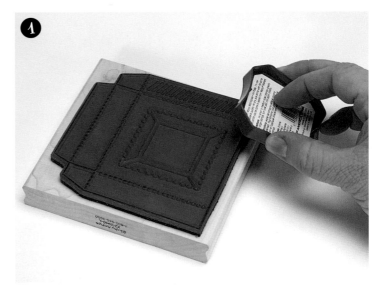

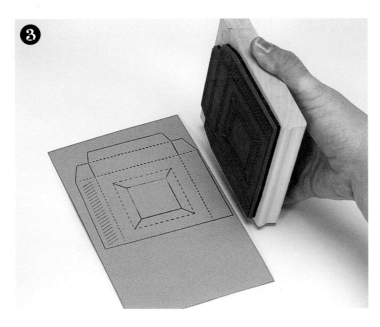

1. Ink Outside of Box Template Stamp
Ink the outside edge of the box template stamp with black dye ink.

2. Stamp on Cardstock
Stamp it onto two separate pieces of green cardstock. One will be the back of the box (panel A) and the other will be the door panel (panel B) that covers the card.

3. Ink and Stamp Inside of Box Template
Ink the entire box template stamp and stamp it onto another piece of green cardstock. This becomes the front of the box that has the picture frame (panel C).

4. Stamp Gift Image

Ink the present stamp and randomly stamp it on the blank side of the panel A with red dye ink. Repeat with panels B and C.

5. Cut Out Box Outline

Cut out the box shapes with craft scissors.

6. Remove Extra Panel

On panel B, cut off the two extra flaps, leaving a single flap to glue on the box.

7. Cut Out Box Center

Cut out the center of panel C with a craft knife.

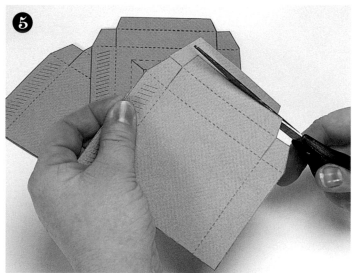

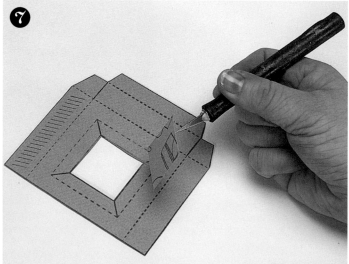

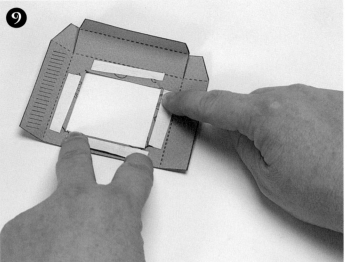

8. Score Along Dotted Lines

Using the back of a craft knife, score along the dotted lines on each panel.

9. Tape Down Frame

Tape down the inside window frame of panel C This gives a much more finished edge to the inside rim of the frame.

10. Add Acetate Square

Cut out a square of clear acetate and adhere it to the inside of panel C with clear-drying glue. Set aside.

11. Ink and Stamp Christmas Tree Image

Ink the Christmas tree stamp with green dye ink and stamp it on white cardstock.

For a stained glass effect,
try using colored acetate!

12. Add Glue
Add dots of clear-drying glue around the tree.

13. Add Glitter "Snow"
Sprinkle the wet glue with a generous amount of glitter to create the look of snow. Wipe away the excess.

14. Trim Cardstock
Trim the white cardstock to fit inside the frame.

15. Assemble Box
Attach panel A to panel C with double-stick tape.

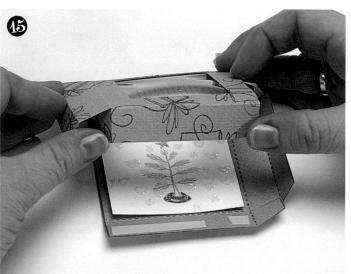

16. Arrange Flaps

Be sure the side flaps on panel C are positioned on the outside of the box.

17. Tape Box Shut

Adhere a strip of double-stick tape to the flap that will be tucked into the box to hold it closed.

NOTE: If you plan to put candy in the box, skip step 17.

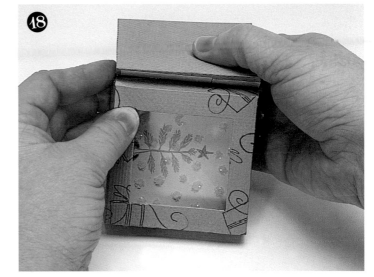

18. Attach Door

Adhere a strip of double-stick tape to the small flap on panel B and attach it to the side of the box.

19. Bend Door to Open and Close

Bend panel B back and forth a few times to make sure it opens and closes properly. What a nice Christmas surprise!

LIGHT THE MENORAH

acrylic paints are great for card-making because they layer well, are lightweight and clean up with water. These pearlized acrylic paints are gorgeous when applied to richly colored paper. For a softer, smoother finish, try using a cosmetic sponge instead of a stiff brush to apply the paint. Be sure the paint is completely dry before embossing so the powder sticks only to the image.

TIP

Edging this card with the same silver color as the embossing powder will really light up this Menorah!

What you'll Need

Stamps*

- Decorative stamp
- Menorah stamp

Materials

- 2 pieces of white cardstock
- Folded silver greeting card
- Dark blue acrylic paint
- Light blue acrylic paint
- Navy blue pigment ink
- Silver pigment ink
- Silver paint pen
- Stiff brush or cosmetic sponge
- Double-stick tape

*See Resource Guide for stamp credits.

1. Apply Dark Blue Acrylic Paint
Generously apply dark blue acrylic paint to a piece of white cardstock with a stiff brush or cosmetic sponge.

2. Apply Light Blue Pigment Paint
Apply light blue acrylic paint to a separate piece of white cardstock.

3. Ink and Stamp Decorative Image
Ink the decorative image with navy blue pigment ink. Stamp along the bottom of the dark blue cardstock.

4. Ink and Stamp Menorah Image

Ink the Menorah stamp with silver pigment ink and stamp it onto the light blue cardstock.

5. Tear Card

Tear the top and bottom of the light blue cardstock to create jagged edges.

6. Outline Cards

Outline the edges of both cards with a silver paint pen.

7. Adhere to Greeting Card

Using double-stick tape, adhere the dark blue layer to the front of a folded silver greeting card. Next, adhere the light blue layer over the dark blue layer.

MORE BRIGHT IDEAS

Butterfly in the Sky

I wanted this majestic butterfly to look like it was emerging from the foliage, so I stamped the butterfly first and used paints and powders to decorate it. Then, I stamped the leaves, with metallic paint, over parts of the butterfly to set it back, behind the leaves.

Leaving an Impression

For this card, I stamped the leaves with dark acrylic paint on the card. Then, I layered a piece of cardstock stamped with lighter acrylic paints over it. The cord and ribbon hide the seam where the two pieces meet.

JOSEPH'S COAT

this festive card is great for ringing in the New Year—or any time of year! If you think back to kindergarten, you might remember using this resist method with crayons and black poster paint. While the end result might look a little more sophisticated, the technique is just as simple.

TIP

The closer and thinner you make the lines, the better this technique looks!

What you'll Need

Stamp*

• Nile flowers stained glass stamp

Materials

• White cardstock
• Silver paper
• Folded white greeting card
• Black solvent ink
• Black dye ink
• Black permanent marker
• Colored pencils
• Craft knife
• Craft scissors
• Craft glue
• Paper towels

*See Resource Guide for stamp credits.

1. Sharpen Colored Pencil
Using a craft knife, sharpen several colored pencils.

2. Ink and Stamp Decorative Image
Ink the Nile flowers stained glass stamp with black solvent ink and stamp onto white cardstock.

3. Color in the Design
Fill in the white spaces of the image with colored pencils. You may color areas solidly or use loose lines to fill in portions of the design like I've done here.

4. Finish Adding Color
Finalize the colored pencil work, adding as much or as little color as you desire.

5. Cover with Dye Ink
Stamp black dye ink over the entire image.

6. Wipe Away Ink
Before the black ink dries, quickly wipe it away with a paper towel. The ink will adhere to all the areas not covered with colored pencil.

7. Trim Cardstock
Trim the sides of the cardstock with craft scissors.

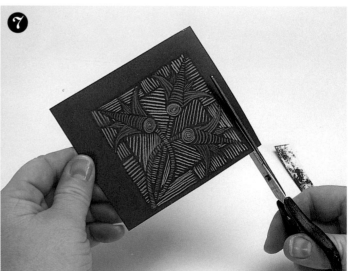

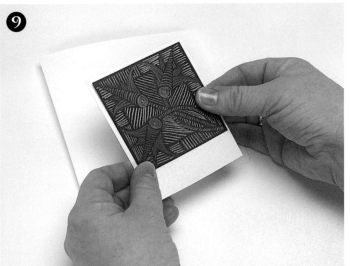

8. Outline Edges

Outline the edges of the cardstock with a black permanent marker.

9. Adhere Cardstock to Paper

Attach the cardstock to a slightly larger piece of silver
paper with craft glue, leaving space at the bottom for a
message if desired.

10. Attach to Greeting Card

Attach the stamped cardstock and silver paper to a
folded white greeting card with craft glue.

GLAZED METALLICS

i have used this technique for many years to achieve a soft metallic finish on paper. It is great to try over an image that did not emboss well or one that is splotchy. High-quality pigment inks and paper make all the difference when crafting your own cards. For best results, I recommend using Encore brand pigment ink.

TIP

The pewter sticker on this card adds a fancy touch. To learn more about pewter stickers, see the Celebration of Summer card on page 76.

What you'll Need

Stamp*

♦ Swirl stamp

Materials

♦ White cardstock
♦ Folded blue greeting card
♦ Pewter sticker
♦ Silver pigment ink (Encore)
♦ Silver embossing powder
♦ Red, green and blue dye
  re-inkers
♦ Embossing tool (heat gun)
♦ Craft scissors
♦ Double-stick tape
♦ Paper towels

*See Resource Guide for stamp credits.

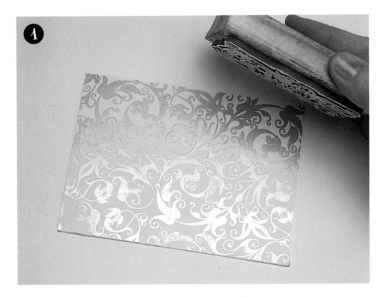

1. Ink and Stamp Decorative Image
Ink the swirl stamp with silver pigment ink (Encore) and stamp onto white cardstock.

2. Add Embossing Powder
Sprinkle silver embossing powder over the ink, then tap the edge of the card to remove the excess powder.

3. Heat-Set Powder
Heat-set the powder with an embossing tool to melt it.

4. Cover Image

Use the same silver stamp pad to stamp all over the cardstock, completely covering the image.

5. Spread Ink with Finger

Rub the silver pigment ink into the surface of the card with your finger.

6. Add Dye

Squeeze a few drops of blue dye from a dye re-inker into a small dish. With your finger, smooth out the ink so it is fairly thin.

7. Buff Image

Buff the image with your inky finger so the dye sticks to the recessed areas. Continue buffing with red and green dye or the colors of your choice.

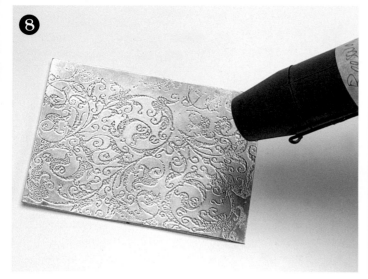

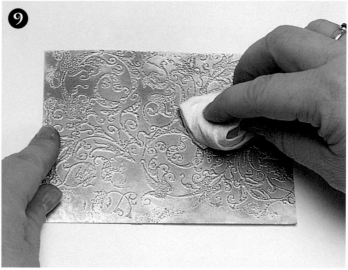

8. Heat-Set Dye

Heat-set the ink and dye with an embossing tool.

9. Buff Cardstock and Adhere to Greeting Card

Buff the cardstock with a paper towel to remove excess ink and dye. Trim the cardstock with craft scissors and adhere to a folded blue greeting card with double-stick tape. Add a pewter sticker for a finishing touch.

 Try this!

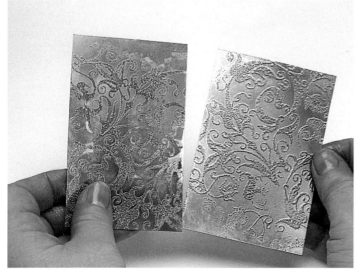

For a rougher look, shown on the left, rub vigorously with a paper towel and/or scrape with a craft knife to remove dye. For a softer look, shown on the right, simply use the buffing technique explained above.

CROSS MY HEART

in February, after the rush of the winter holidays, I am ready to create a few fun valentines with this tricky old pop-up inside. Be sure to add lots of confetti to the cards and send them in brightly colored envelopes. (Sealed with a kiss, of course.)

TIP

Be careful not to go overboard on your colors. For a simple and visually appealing look, I recommend choosing no more than three colors.

What you'll Need

Stamps*

- Hand-carved heart stamp (template on page 44)
- "Happy Valentine's Day" stamp

Materials

- Heart pocket template (page 45)
- Translucent white vellum
- Folded white greeting card
- Bright construction paper
- Dye ink, various colors
- Pencil
- Eraser
- Carving tools
- Double-stick tape
- Decorative paper punch
- Stylus
- Craft scissors or craft knife

See Resource Guide for stamp credits.

1. Cut and Trace Heart Pocket

Cut out and enlarge the heart pocket template provided on page 45. Trace the outline and score marks of the template onto a piece of translucent vellum with a pencil.

2. Score Template

Score inside the slits with a stylus. Cut out the template with craft scissors or a craft knife.

3. Clean Up Vellum

Erase any pencil marks remaining on the vellum.

4. Crease Vellum
Crease the vellum along the score marks to make the heart pocket. The creases down the center of each heart should fold to the inside.

5. Stamp Images on Vellum
Carve a heart stamp using the template provided below (For carving instructions, see page 17). Randomly stamp the heart and "Happy Valentine's Day" stamps onto the inside of the heart pocket in various colors of dye ink.

6. Stamp Heart Image on Greeting Card
Stamp more hearts onto the outside of a folded white greeting card in various colors.

7. Position Heart Pocket in Greeting Card
Fold the heart pocket back into shape and place it inside the greeting card. Try closing the card to make sure it is in the correct position.

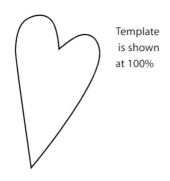

Template is shown at 100%

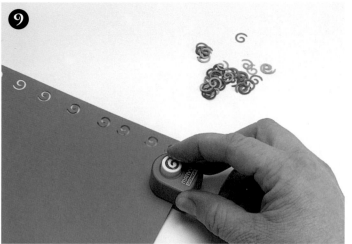

8. Adhere Heart Pocket
Adhere the heart pocket to the inside of the white greeting card with double-stick tape.

9. Punch Out Confetti
Use a decorative paper punch on bright construction paper to make colorful confetti.

10. Add Confetti
Open the card slightly, pour in the confetti, and carefully re-close the card.

Enlarge heart pocket template by 200%, then by 125%.

TIP

Add glitter to the inside
of the heart for a shiny
(and messy) touch.

HEARTS OF GOLD

this is a creative method of making vivid backgrounds for simple stamps. With a few dye inkpads and a stiff brush for applying color, you can count on getting brilliant results every time! The torn-edge background adds a modern flair and works exceptionally well for underwater or sunset themes.

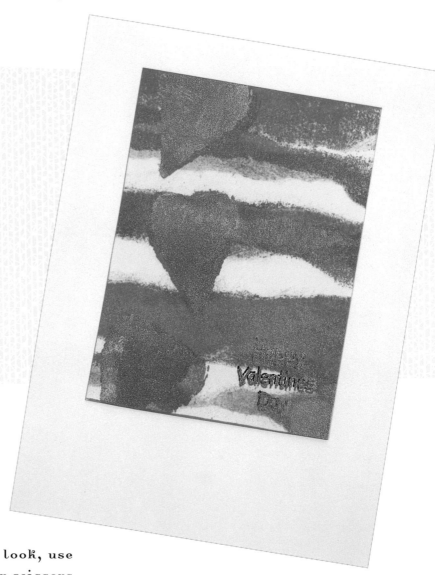

TIP

For a more uniform look, use decorative or regular scissors to cut the Eclipse tape.

What you'll Need

Stamps*

- Heart stamp
- "Happy Valentine's Day" stamp

Materials

- White cardstock
- Folded white greeting card
- Dye ink, various colors
- Gold pigment powder
- Eclipse tape
- Stiff brush
- Craft scissors

See Resource Guide for stamp credits.

1. Mask Background
Tear off thin strips of Eclipse tape and adhere them to a piece of white cardstock to mask portions of the background.

2. Apply Dye Ink
Brush colored dye ink onto the cardstock and over the masks with a stiff brush. Remove and reposition the masks until the desired look is achieved.

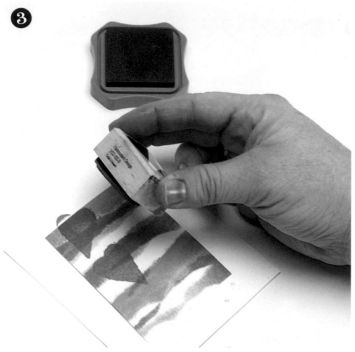

3. Ink and Stamp Heart Image

Ink the heart stamp with pink dye ink and stamp randomly onto the cardstock.

4. Ink and Stamp "Happy Valentine's Day"

Ink the "Happy Valentine's Day" stamp with black dye ink and stamp it onto the bottom right corner.

5. Add Pigment Powder and Adhere to Greeting Card

Dust with gold pigment powder and wipe away the excess. The powder will stick to the wet ink on the hearts and text. Trim the edges and adhere to a white greeting card.

MORE BRIGHT IDEAS

Tie-Dyed Flowers

Torn edges add dimension to cards such as this colorful creation. Hold a piece of paper (cardstock works best) against a flat surface with one hand and slowly tear the paper with the other. Voila!

Midas' Golden Touch

The shimmer of gold cardstock shines through on this card. Don't worry about tearing the edges in a straight line. The more jagged the tear, the better.

CARDS FOR spring

after a long winter, the bright colors of spring are like a breath of fresh air! These spring cards are filled with the vibrant colors of a season in bloom. Flowers and nature themes always make great cards for Easter and Mother's Day, but don't forget to keep your eyes open for fun and festive Mardi Gras imagery as well.

party gras!

these days, you don't have to go to New Orleans to celebrate Mardi Gras. This extravaganza of crazy costumes and delectable delights is an excellent time to create a special card, no matter where you happen to live. The mix of creative patterns and vivid colors will whisk you away to The Big Easy, where you'll toss your troubles aside and just have fun!

TIP

To decorate your card in true Mardi Gras fashion, use the official colors of the festival: purple represents justice; green represents faith; and gold represents power.

What you'll Need

Stamps*

- Box template stamp
- Minstrel stamp

Materials

- 2 sheets of decorative paper
- White paper
- Acetate square
- Folded dark blue greeting card
- Black solvent ink
- Silver paint pen
- Colored pencils
- Glitter
- Double-stick tape
- Scoring tool
- Craft scissors or craft knife
- Ruler

See Resource Guide for stamp credits.

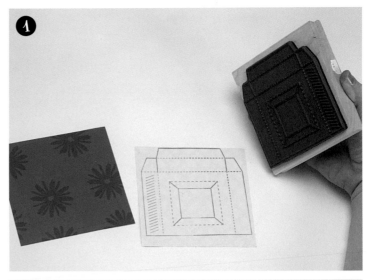

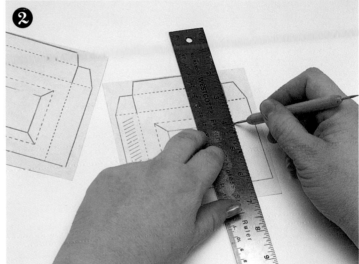

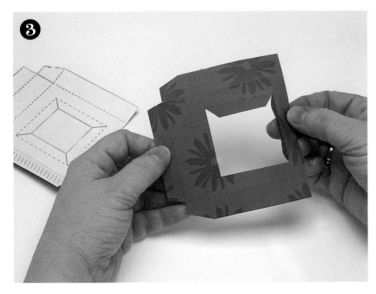

1. Cut and Stamp Decorative Paper

Cut two thin pieces of decorative paper to approximately 3" (7.6cm) square. Stamp the plain sides of each piece of paper with the box template stamp.

2. Score Paper

On one of the pieces of paper, score the outside lines and the window of the box with a stylus. This will be the frame. On the other piece of paper, score only the outside of the box. This will be the back panel for the frame. Cut out both shapes with scissors or a craft knife.

3. Fold Box

Fold the frame along the score marks.

4. Remove Top Flaps
Cut off the little box flaps and bottom flap of the frame with craft scissors.

5. Remove Back Flaps
Trim off all flaps but one from the back panel piece.

6. Remove Top of Flap
Cut the top portion off of the remaining flap on the back panel piece, leaving a single tab.

7. Add Acetate Window
Cut a small square of acetate approximately 2½" x 2½" (6.4cm x 6.4cm) for the frame's window. Adhere the acetate with double-stick tape on the inside of the frame. Attach the remaining back panel flap to the inside of the frame as shown. Be sure the decorative paper is showing through the frame.

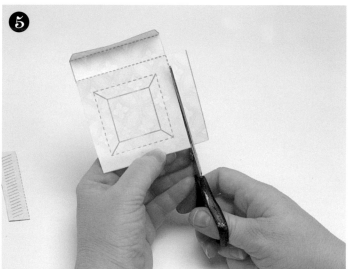

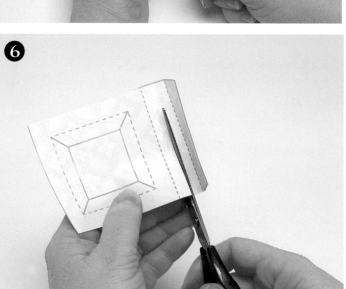

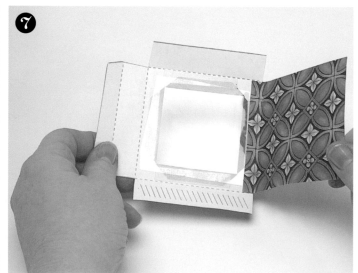

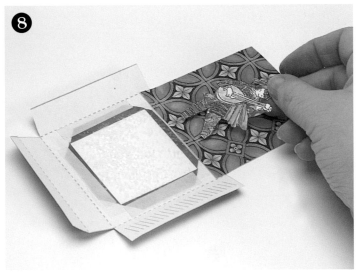

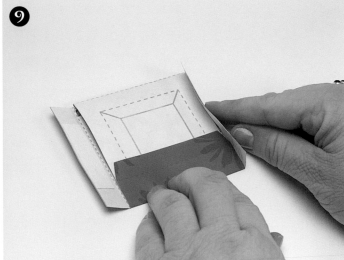

8. Add Minstrel Image

Stamp the minstrel image onto white paper using black solvent ink. Cut it out and decorate it with colored pencils. Place the image in the center of the back panel and add glitter to the window. Fold the frame over the back panel.

9. Fold Flaps

Fold the remaining flaps from the frame in and secure with double-stick tape. Edge the frame with a silver paint pen.

10. Adhere Box to Greeting Card

Use double-stick tape to fasten the decorative framed piece a folded dark blue greeting card.

PEEKABOO BUNNY

Whimsy meets whiskers in this Easter card, made just for a special little girl or boy. Tuck it into an Easter basket along with some colored eggs. To take this card a step further, glue Easter grass to the bottom of the card and add a small pink button for the bunny's nose.

TIP

This tri-fold method can also be used for other cards, so don't limit yourself to this design alone.

What you'll Need

Stamps*

- Swirl stamp
- Hand-carved Easter egg stamp (template on page 61)
- Hand-carved bunny face stamps (template on page 61)

Materials

- Bunny body template (page 61)
- Light yellow cardstock
- Vellum or tracing paper
- Blue solvent ink
- Dye ink, various colors
- Pencil
- Eraser
- Carving tools
- Craft knife
- Craft scissors
- Ruler

*See Resource Guide for stamp credits.

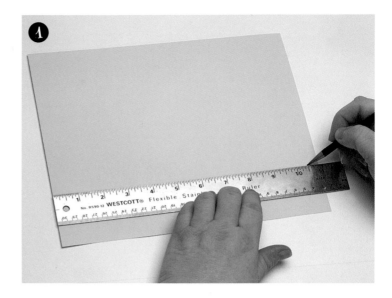

1. Mark Cardstock

Measure 3" (7.6cm), 7" (18cm) and 10" (25cm) and mark with a pencil on light yellow cardstock.

2. Score Cardstock

Score along the 3" (7.6cm) and 7" (18cm) marks with the back of a craft knife or a scoring tool. Erase any remaining pencil marks.

3. Tri-Fold Card

Fold the left panel to the back and the right panel to the front. This is called a tri-fold.

4. Ink and Stamp Swirls

Ink a swirl stamp with blue solvent ink. Flatten out the card and begin stamping the front about one-third of the way up on the left edge. Work your way upward and diagonally to the right edge of the page.

5. Fill Open Spaces

Re-ink the stamp and randomly stamp in between the open spaces of the yellow cardstock to fill in any gaps.

6. Trim Card

Use your craft knife to trim off the top of the card along the stamped edge.

7. Stamp Back of Card

Flip the card over and stamp randomly along the top edges with the swirl stamp.

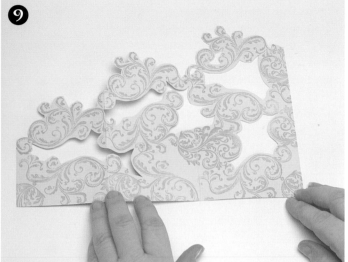

8. Fill Open Spaces
Fill the open spaces on the edges with more stamping.

9. Cut Out Random Pieces
Use your craft knife to cut out random bits and pieces from the inside of the card.

10. Trim Card
Using your ruler as a guide, trim the edges of the card with a craft knife.

11. Ink and Stamp Easter Egg Image
Carve an Easter egg stamp using the template on page 61. (For carving instructions, see page 17). Ink the stamp with colorful dye ink and stamp a border along the bottom of both sides.

Many Easter egg stamps can be found in your local stamping supply store if you don't wish to carve your own.

12. Cut Out Bunny Template

Cut out the bunny template provided on page 61.
Position the straight side of the template against a
folded piece of vellum or tracing paper and trace with
a pencil. Cut out the bunny shape.

13. Add Bunny Eyes and Nose

Dip a pencil eraser in pink dye ink and use it to dot the
bunny's eyes. Use a small triangular piece of eraser to
make a nose, and stamp it with pink dye ink.

14. Add Bunny Ears

Use the template on page 61 to carve the ear shape
from an eraser. Stamp it inside each of the bunny's ears
with pink dye ink. Poke the bunny's head through
holes in the back of the card.

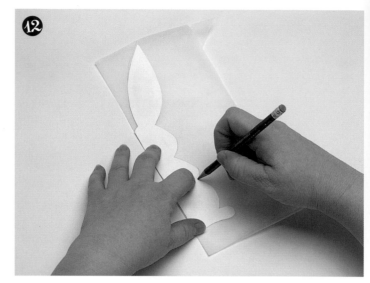

TEMPLATES

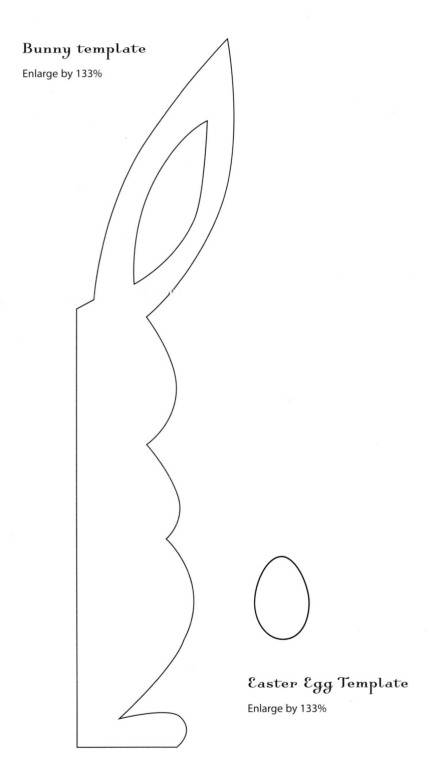

Bunny template

Enlarge by 133%

Easter Egg Template

Enlarge by 133%

sunbonnets in spring

here is a favorite trick of mine! The lamp stamps I used here don't have anything to do with spring, but with a little imagination and clever masking, they're transformed into bonnets! Take a good look at your stamps and try to imagine what else they could be. Who knows what you'll dream up!

TIP

This process works better when using different kinds of ink, or at least drastically different colors.

What you'll Need

Stamps*

- Dress stamp
- Chinese lantern stamp
- Lamps stamp

Materials

- White paper
- White cardstock
- Folded white greeting card
- Dye ink, various colors
- Silver dye ink
- Turquoise paint pen
- Eclipse tape
- Double-stick tape
- Craft scissors

*See Resource Guide for stamp credits.

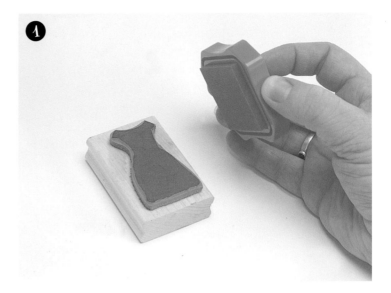

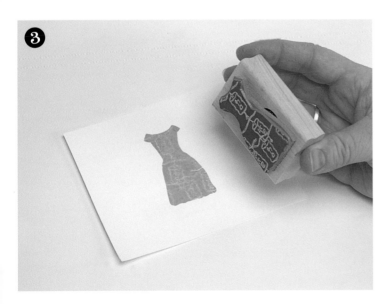

1. Ink Dress Image
Ink the surface of the dress stamp generously with orange dye ink.

2. Ink and Stamp Lantern Image
Ink the Chinese lantern stamp with silver dye ink and stamp the image onto the solid dress stamp to create designs on the dress.

3. Stamp Inked Dress Image
Stamp the dress image onto white paper.

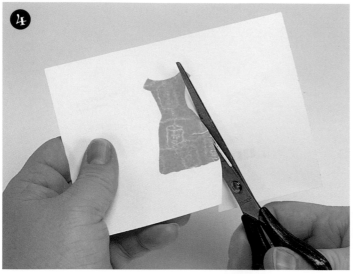

4. Cut Out Dresses
Cut out the dress image with craft scissors. Repeat this process two more times using different colors.

5. Ink and Stamp Lamp Image
Ink the lamp stamp with various brightly colored dye inks and stamp the top of a piece of white cardstock.

6. Mask and Stamp Lamp Bases
Mask off the center of the white cardstock with Eclipse tape and stamp the cardstock again with the lamp image. Only the lamp bases should show up on the cardstock.

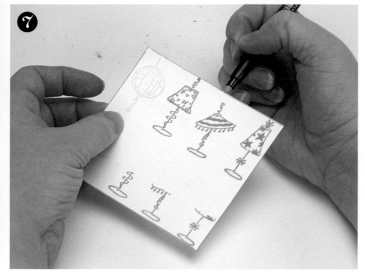

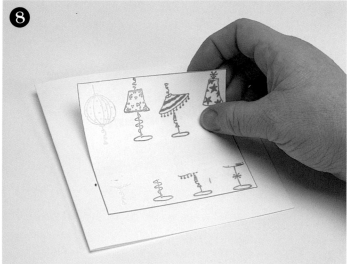

7. Trim Cardstock and Outline Edges

Trim the sides of the cardstock to make it square.
Outline the edges with a turquoise paint pen.

8. Adhere Cardstock to Greeting Card

Fasten the square cardstock onto a folded white greeting card with double-stick tape.

9. Adhere Dresses Over Lamp Bases

Adhere the cut-out dresses over the lamp bottoms
with double-stick tape.

A MOTHER'S LOVE

mom always appreciates a beautiful card—particularly one made by you! Creating a special card just for her will give you the chance to express your love in a way a store-bought card never could. Use her favorite colors and themes. Add a personal embellishment or an old photograph she will recognize. This is your opportunity to show your mom how much you have appreciated her all these years.

TIP

Galaxy markers are great for coloring in stamped images on light or dark paper. They are available in bright and pastel colors, metallics and glitters, and with fine, medium and broad tips.

What you'll Need

Stamp*

- Triple heart stamp

Materials

- Black solvent ink
- Pink dye ink
- Markers (Galaxy)
- Folded white greeting card

See Resource Guide for stamp credits.

1. Ink and Stamp Heart Image

Ink the triple heart stamp with black solvent ink and stamp it onto the front of a folded white greeting card.

2. Decorate Hearts

Color in the hearts with markers (I use Galaxy brand) and allow the ink to dry.

3. Stamp Over Hearts

Use a pink dye inkpad to stamp over the triple heart image in the opposite direction.

CARDS FOR SUMMER

Summer cards are the forgotten greetings. Between vacations, soccer practice and swimming lessons, who has time for another project? But that doesn't mean there aren't many special events to celebrate, like Graduation, Father's Day and Summer Solstice. Card-making can be done virtually anywhere! A blanket, a cool glass of lemonade and your card supplies are all you need.

FACING THE FUTURE

g raduation is one of life's most cherished moments. Show your favorite grad how much he means to you with a special handmade greeting card. This one can be customized with the grad's school colors. Be sure to leave space for a personalized message (and perhaps a little spending money!).

TIP

Fill in the blank spaces of stamped images to give your cards a more polished look. I use thicker Galaxy markers to color in larger areas and gel pens for smaller, more detailed areas.

What you'll Need

Stamps*

- Large frame stamp
- Small frame stamp
- David stamp

Materials

- Blue cardstock
- Folded blue greeting card
- Black solvent ink
- White marker
- White gel pen
- Double-stick dots
- Craft scissors
- Craft knife

*See Resource Guide for stamp credits.

1. Stamp Frame Images
Cut the blue cardstock into two pieces, a larger one measuring about 4" x 6" (10.2cm x 15.2cm) and a smaller one measuring about 3" x 3½" (7.6cm x 8.9cm). Using black solvent ink, stamp the larger card with the large frame stamp and the smaller card with the small frame stamp.

2. Color in Open Spaces
Fill in all open spaces of the stamped frames with a white marker (I used Galaxy brand) and gel pen.

3. Trim Card Edges
Trim around the edges of both frames with craft scissors.

4. Cut Out Window
Cut out the inside window of the small frame with a craft knife.

5. Stamp and Color David Image
Stamp the David image onto a separate piece of blue cardstock and color in larger areas with a white marker.

6. Fill in Open Spaces
Fill in smaller areas with a white gel pen. For added detail, make tiny dots of gel pen on the face.

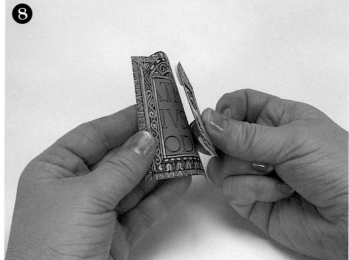

7. Adhere Small Frame to Face

Position the small frame over the David image so most of it can be seen through the window, and adhere with double-stick dots. Trim away all of the excess around the face.

8. Adhere Small Frame to Large Frame

Adhere the David image and small frame to the front of the large frame with double-stick dots.

9. Add Dots

Place more double-stick dots on the back of the frame.

NOTE: To adhere small areas, try cutting double-stick dots into smaller pieces with craft scissors.

10. Adhere to Greeting Card

Fasten the framed face to the front of a folded blue greeting card, leaving room at the bottom for a special message if desired.

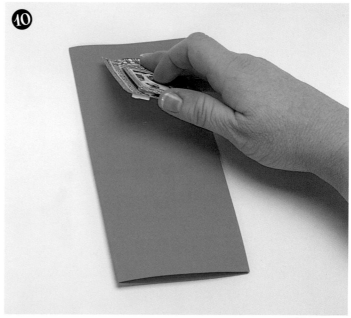

IN THE MAIL

blank postage paper is available at many stamp and scrap-book stores. Or, try making your own by using a sewing machine without thread to perforate the paper. I love this little big head stamp. I think it's perfect for Father's Day!

What you'll Need

Stamp*

- Little big head stamp

Materials

- Perforated stamp paper
- Folded dark blue greeting card
- Black dye ink
- Dye ink, various bright colors
- Markers
- Gel pens
- Eclipse tape
- Stiff brush
- Double-stick tape
- Craft knife

See Resource Guide for stamp credits.

TIP

With simple masking, create a thin white border around the image for a professional postage look.

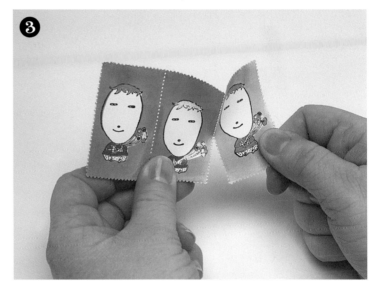

1. Ink and Stamp Image

Ink the stamp with black dye ink and stamp several times onto perforated stamp paper. Stamp the image onto Eclipse tape, and use a craft knife to cut out the outline of the image to make a mask. Apply the mask over three of the stamped images on the postal stamp.

2. Apply Dye

Apply various colors of dye ink with a stiff brush. Next, remove the mask and color in the stamped images with markers (I used Galaxy brand) and gel pens.

3. Separate Stamps

Tear the stamp paper apart at the perforations.

4. Adhere to Greeting Card

Cut pieces of double-stick tape to fit the back of each stamp and adhere diagonally from top-left to bottom-right onto the folded dark blue greeting card.

CELEBRATION OF SUMMER

i love these real pewter stickers, which are fairly new on the market. They come in a variety of shapes and designs. For this card, I uses inks to color in the fancy embellishments, but you may also prefer the pewter's natural finish. The 100% pewter stickers are soft and bendable, and they can be cut to size with craft scissors. Why not have a family reunion, summer solstice celebration or Independence day grill-out? The beauty of this card is that you can color coordinate it for any occasion.

TIP

This technique also works with foil stickers.

What you'll Need

Materials

- 6 pieces of decorative cardstock
- Folded tan greeting card
- Red and green solvent ink
- 2" (5cm) square pewter sticker
- Cosmetic sponge
- Craft scissors
- Double-stick tape
- Paper towels

1. Cut Out Decorative Cardstock

Cut six pieces of decorative cardstock into squares of descending size. The sizes in this project are as follows: 5" x 5" (13cm x 13cm), 4½" x 4½" (11cm x 11cm), 4" x 4" (10cm x 10cm), 3½" x 3½" (9cm x 9cm), 3" x 3" (8cm x 8cm), 2⅛" x 2⅛" (7cm x 7cm). The pewter sticker should measure 2" x 2" (5cm x 5cm), and the greeting card should measure 5½" x 5½" (14cm x14cm). Layer the decorative cardstock in descending order with the smallest square on top. Adhere all the layers together with double-stick tape.

2. Ink Pewter Sticker

Apply two colors of solvent ink to the pewter sticker. For this project, I used red and green.

3. Spread Ink

Spread the ink around with a cosmetic sponge, completely covering the sticker.

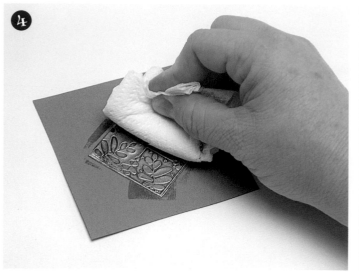 

4. Remove Excess Ink
Blot the excess ink from the pewter stamp with a paper towel. Allow the ink to dry.

5. Adhere to Greeting Card
Using double-stick tape, adhere the layered cardstock to the greeting card. Next, adhere the pewter sticker to the cardstock.

MORE BRIGHT IDEAS

All the Leaves are Brown

Overlapping leaf stickers create the background on the pewter sticker in this card. Then, I used individual stickers in the center and along the edge as an accent. The pieces of the card are layered using double-stick dots to give it even more impact.

Toilettes

I applied a generous amount of dark green and blue solvent ink to the pewter sticker on this card. As a further embellishment, I also added a black satin ribbon trimmed in gold.

Daisies and Checks

Pewter stickers come in all shapes and sizes, as you can see in this card. Scrapbook stores have all sorts of coordinated papers where you can get pieces like these checks and stripes.

CARDS FOR FALL

fall is my favorite time of year. I love when the temperature drops and leaves turn beautiful shades of orange, brown and gold. It's the perfect time to roll up the sleeves of your favorite sweatshirt and get crafting! With a little ink, some colorful paper and a few stamps, you'll be amazed at how time flies. You may even get a jump on your winter holiday cards.

SPOOKY SENTIMENTS

halloween is my all-time favorite holiday! It is such a fun time, purely for tricks and treats (my friends know I'm a real joker). It's thrilling to create cards for this spooky holiday because they can be scary, humorous, cute or a mixture of all three!

TIP

To prevent your embossing powder from settling and sticking to the bottom of the jar, give it a shake before and after every use.

What you'll Need

Stamps*

- Shrine stamp
- Pumpkin head stamp

Materials

- Orange cardstock
- Black cardstock
- Black pigment ink
- Black permanent marker
- Silver pigment ink
- Embossing powder
- Embossing tool (heat gun)
- Markers
- Gel pens
- Double-stick dots
- Craft knife

*See Resource Guide for stamp credits.

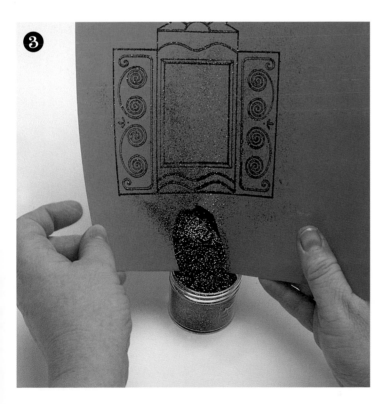

1. Ink and Stamp Frame Image

Ink the shrine stamp with black pigment ink and stamp onto orange cardstock. Quickly touch up the stamped image with a black permanent marker.

2. Apply Embossing Powder

Pour a generous amount of embossing powder over the stamped image. The powder will stick to the wet ink.

3. Save Excess Powder

Return the excess powder to the container.

4. Heat-Set Powder

Heat the stamped image with an embossing tool to set the powder in place.

5. Cut Out Center of Frame

Cut out the center of the shrine image with a craft knife. This becomes the frame for the card.

6. Color Frame

Color in portions of the frame with markers (I used Galaxy markers) and gel pens.

7. Ink and Stamp Halloween Image

Ink the pumpkin head stamp with silver pigment ink and stamp it onto black cardstock.

8. Adhere Image

Place the frame over the pumpkin head image and adhere with double-stick dots.

9. Make Stand

To make a stand for the frame, cut a triangle out of a scrap piece of black cardstock. Affix it to the back of the frame with double-stick dots.

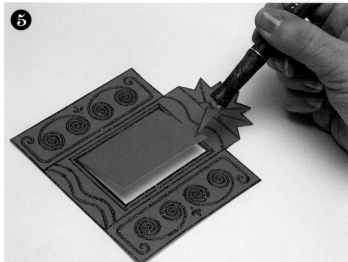

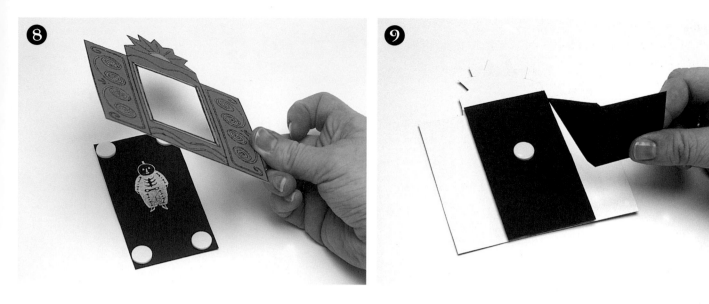

MORE BRIGHT IDEAS

Pumpkin Patch

This is another variation on the Halloween theme. I used double-stick dots to adhere the layers and to make the card three-dimensional.

GOBBLER GREETINGS

in this project, you get to carve your own turkey—turkey stamp, that is. Use the template provided to cut out the cute little gobbler. But beware! Carving can be addictive, and creating simple shapes from erasers can broaden your stamp collection. You'll quickly find that extra storage space in your craft room is something to be thankful for!

TIP

To deepen the warm fall tones in this card, I used an all-over vivid yellow paint as an overlay. This technique can be used with other colors as well. For instance, pink can be applied over blue and purple to brighten the look.

What you'll Need

Stamps*

- Hand-card turkey stamp (template on page 89)
- Leaf stamp

Materials

- Cream-colored cardstock
- Burgundy cardstock
- Folded tan greeting card
- Dye ink, various bright colors
- Black permanent marker
- Cosmetic sponge or stiff brush
- Pencil
- Eraser
- Carving tools
- Double-stick tape
- Craft scissors

*See Resource Guide for stamp credits.

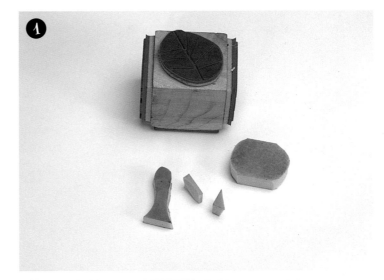

1. Carve Turkey Stamp
Carve the individual pieces of the turkey stamp using the templates provided on page 89. (For instructions on how to carve your our own stamps, see page 17).

2. Ink and Stamp Turkey Feathers
Ink a leaf stamp with several colors of dye ink and stamp onto cream-colored cardstock to form the turkey's feathers.

3. Stamp Turkey Body
Ink the body stamp with brown dye ink and stamp over the feathers.

4. Stamp Turkey Head
Ink the turkey's head stamp with brown dye ink and stamp above the body.

5. Stamp Turkey Beak and Feet
Ink the turkey's beak and feet stamps with orange dye ink and stamp below the body.

6. Stamp Turkey Legs
Repeat the process with the turkey's legs.

7. Dot Turkey Eye
Dot the turkey's eye with a black permanent marker.

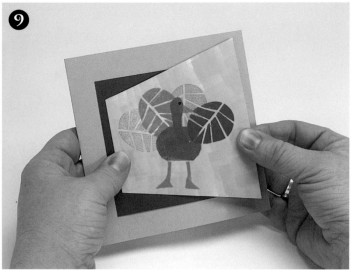

Templates are shown at 100%.

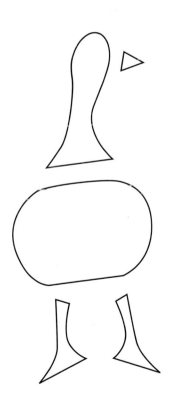

8. Cover with Yellow Ink

Cut the cardstock into an interesting geometric shape with craft scissors. Spread yellow dye ink over the entire turkey image with a cosmetic sponge.

9. Adhere Turkey Cardstock

Cut a piece of burgundy cardstock to the desired size and adhere to a folded greeting card with double-stick tape. Adhere the turkey card over the burgundy cardstock.

FALL FOLIAGE

acrylic tiles come in a myriad of shapes and sizes and can be found in most stamp supply stores. The clear tiles are lightweight enough to be used on cards. The ink selection is important on these tiles. Be sure to use a solvent ink or a pigment that can be dried with a heat gun. You may also want to try embossing the tiles for a metallic look.

What you'll Need

Stamp*

- Leaf stamp

Materials

- Thin decorative paper
- Folded light green greeting card
- Acrylic tile
- Metallic paint
- Copper paint pen
- Black solvent ink
- Stiff brush
- Double-stick tape
- Craft scissors

See Resource Guide for stamp credits.

1. Paint Acrylic Tile

Use a stiff brush to cover an acrylic tile completely with metallic paint. (I used Halo Pink Gold paint by Lumiere.)

2. Apply Paint Pen

Fill in the unpainted areas with a copper paint pen. Allow to dry.

3. Remove Backing

Turn the acrylic tile onto the unpainted side and remove backing.

4. Ink and Stamp Leaf Image

Ink the leaf stamp with black solvent ink and randomly stamp it onto the unpainted side of the acrylic tile.

5. Reveal the Image

Continue stamping the leaves as desired. Allow to dry. The metallic paint shows through the stamped, unpainted side of the tile brilliantly.

6. Affix Decorative Paper

Using double-stick tape, adhere a piece of thin, decorative paper to the front of a light green greeting card at the fold. Trim off the excess.

7. Adhere Acrylic Tile

Adhere the acrylic tile to the front of the greeting card over the decorative paper.

Option: Acrylic Tile Experiment

There are so many creative ways to use acrylic tiles for craft projects. Here's a quick and easy idea I think you'll enjoy. The frog-shaped acrylic tile is covered with delicate gold leaf, giving it a faux metallic finish. Once the tile is dry, add it to any greeting card for instant shine.

1. Remove Tile Backing
Remove the backing from the frog-shaped acrylic tile.

2. Add Glaze
Using your finger, spread adhesive glaze on the frog piece.

3. Adhere Tile to Gold Leaf
Stick the adhesive side of the frog piece to gold leaf and allow to dry. (I used foil in this project, but tissue paper works just as well.)

4. Remove Excess Gold Leaf
Peel away the gold leaf and brush off any excess with a soft brush.

5. Fill with Paint Pen
Fill in any spaces with a gold paint pen.

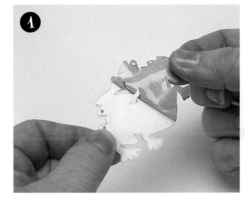

Use a full, soft brush to dust off the excess foil. A brush that is too stiff may remove too much foil and leave a scratchy appearance.

CARDS FOR TIMELESS OCCASIONS

many special occasions celebrated throughout the year are not marked by a particular season. Birthdays, anniversaries and housewarming parties are just a few of the many events that will keep you busy making cards all year long. Add a personal touch by decorating an animal-lover's card with a paw-print stamp, or embellish a history buff's card with vintage images. If you can dream it, you can make it. That, to me, is by far the best part about card-making.

TINY DANCER

this is a great card for a girly-girl party, be it a birthday, shower or other special occasion. Pink and white are the ultimate feminine color combination, although any pastel colors will work just as well.

TIP

Use Eclipse tape to mask off the right side of the card in an uneven line. This will give the illusion of torn paper. This technique is also used in the "A Mother's Love" card on page 66.

What you'll Need

Stamps*

- Harlequin pattern stamp
- Ballerina stamp

Materials

- Folded white greeting card
- Light pink and dark pink pigment ink
- Black solvent ink
- Eclipse tape
- Light pink ribbon
- Stiff brush
- Craft scissors

*See Resource Guide for stamp credits.

1. Mask Greeting Card
Use torn Eclipse tape to mask the outer edges of a folded white greeting card. Apply a generous amount of light pink pigment ink to the card.

2. Texturize Ink
While the ink is still wet, rub a stiff brush over the ink to texturize it. Allow the card to dry completely.

3. Add More Color
Repeat the process, applying dark pink pigment ink over the light pink layer.

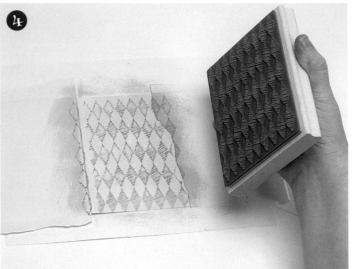

4. Ink and Stamp Harlequin Image
Ink a large harlequin patterned stamp with dark pink pigment ink and stamp over the greeting card.

5. Soften Image
To create a soft look, smear the image with your finger.

6. Ink and Stamp Ballerina Image
Ink a ballerina stamp with black solvent ink and stamp over the harlequin pattern.

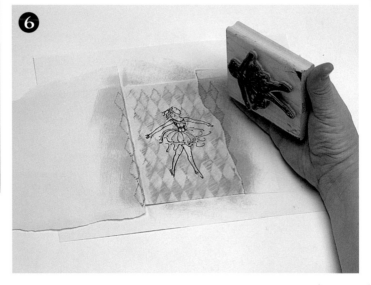

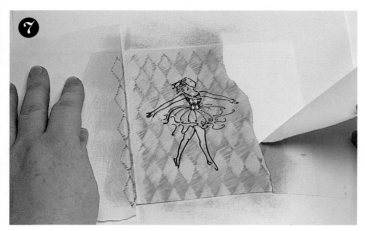

7. Remove Mask

Remove the mask of Eclipse tape and allow to dry.

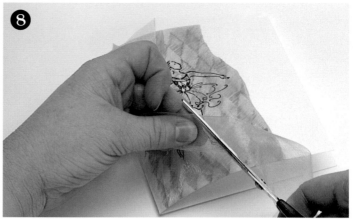

8. Add Ribbon

Fasten a light pink ribbon to the front of the card and tie it in a bow. Cut the ribbon edges at an angle for a decorative look.

More Bright Ideas

Marshmallow Bunny

Before the decorative images were stamped onto the background, masks were applied to the marshmallow bunny image and the right side of the card. When the masks were removed, these areas remained unstamped. This creates a layered appearance.

I DO

When we think of wedding cards, we typically visualize the traditional white-on-white color scheme. But what about the nontraditional couple? Colorful weddings are becoming more and more popular today, and a splash of color will make an ordinary card bright and beautiful—just like the happy couple!

TIP

To make a fancy greeting card like this one, paint a piece of white card-stock with a sandy paint. Then, add an iridescent interference paint over the sandy paint to pearlize it. Add more interference paint to pearlize again. Score the card down the middle with a stylus or the back of a craft knife and apply the stamped image as desired.

What you'll Need

Stamps*

- Bride and groom stamp

Materials

- White cardstock
- Folded iridescent greeting card (see tip on page 100)
- Black dye ink
- Pearlized paint
- Silver paint pen
- Glitter
- Decorative ribbon
- Pewter sticker
- Double-stick tape
- Craft scissors
- Cosmetic sponge
- Paintbrush
- Brayer (optional)

*See Resource Guide for stamp credits.

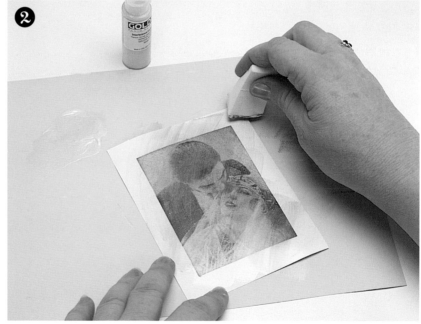

1. Ink and Stamp Bride and Groom Image

Ink the bride and groom image with black dye ink. Stamp onto a piece of white cardstock.

NOTE: When inking a large stamp such as this, use a brayer to distribute the ink more evenly.

2. Apply Paint

Spread pearlized paint over the stamped image with a cosmetic sponge.

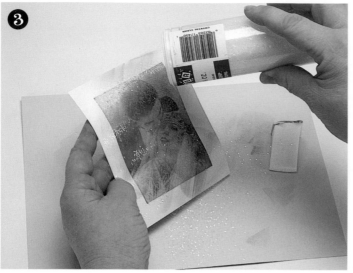

3. Apply Glitter
Add glitter to the wet paint and dust off the excess. Allow to dry.

4. Outline Edges
Trim the edges of the cardstock right up to the edge of the stamped image. Outline the edges with a silver paint pen.

5. Add Ribbon and Adhere to Greeting Card
With double-stick tape, adhere the cardstock and a decorative ribbon to a folded silver greeting card. Finally, adhere a pewter sticker over the ribbon.

More Bright Ideas

Family Matters

This card is an example of how glitter can be a subtle embellishment. Adhere the image to a glossy greeting card in a soft shade of turquoise, and attach a decorative ribbon for a charming touch.

Stars in Your Eyes

A simple, white greeting becomes a sparkling masterpiece just by adding glitter to a simple face stamp. Black ribbon and decorative paper also add visual appeal.

BUNDLE OF JOY

Pastel pink or baby blue, a handmade baby shower invitation will surely make mommy and her guests smile. With the help of a box template, this quick and easy card looks like it took all day to make. I used a simple envelope template to make an all-in-one look. It can be used for other occasions, too. Make several of these at a time to keep ahead of your card-making for every occasion.

TIP

Use yellow cardstock and green ink for a gender-neutral invitation.

What you'll Need

Stamps*

◆ Baby bottle stamp
◆ "You're Invited" stamp
◆ Little gift stamp

Materials

◆ Box template (page 107)
◆ Pink or blue cardstock
◆ White cardstock
◆ Dye inks, various colors
◆ Pink or blue paint pen
◆ Pencil
◆ Stylus
◆ Double-stick tape
◆ Craft scissors

See Resource Guide for stamp credits.

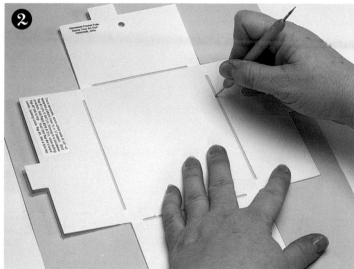

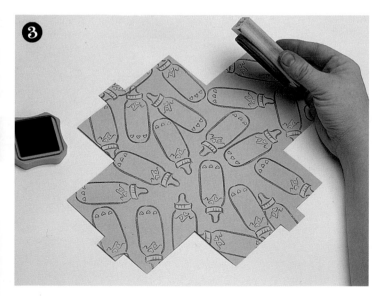

1. Cut Out Template

Enlarge and cut out the box template on page 107. Trace the template with a pencil on the pink or blue cardstock.

2. Score and Cut Box

Score with a stylus where the folds will be, and cut out the box with craft scissors.

3. Ink and Stamp Bottle Image

Ink the baby bottle stamp with colored dye ink and stamp randomly onto the box. Set aside.

4. Decorate Inside of Card

Cut out a piece of white cardstock to the size of the inside of the box, about 5½" x 4¼" (14cm x 10.8cm). Ink the "You're Invited," baby bottle and little gift stamps with various colored dye inks and stamp onto the white cardstock. Outline the edges with a pink or blue paint pen.

5. Adhere Cardstock to Inside of Box

Adhere the white cardstock to the inside back panel of the box with double-stick tape.

6. Close Box

To close the box, fold the flaps inward beginning with the top and bottom, then the left and right. Tuck the flaps under to secure.

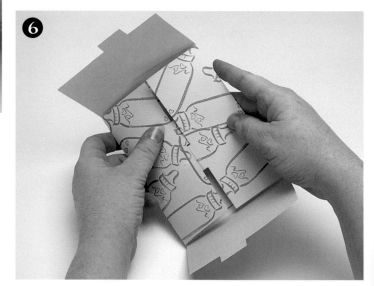

Enlarge template to 167%.

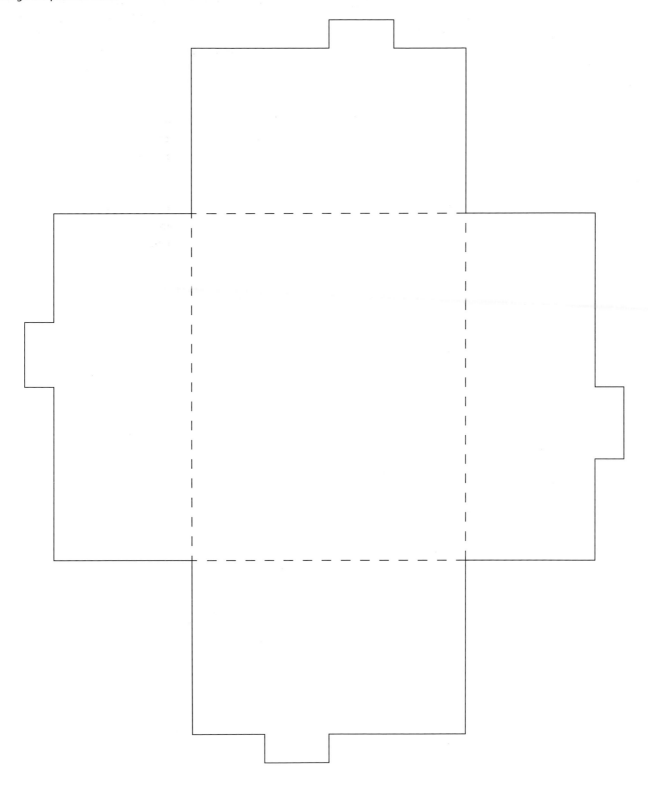

SWEET CELEBRATION

translucent vellum is still the easiest way to add elegance to any card. The variety of colors and patterns available in most craft stores makes it easy to create the ultimate individual look. Choose colors and prints that remind you of the birthday girl or boy and find wrapping paper to coordinate!

What you'll Need

Stamps*

- "Happy Birthday" stamp
- Birthday cake stamp

Materials

- 2 pieces of decorative paper
- Turquoise vellum
- Folded white greeting card
- Silver dye ink
- Purple dye ink
- Double-stick tape
- Craft scissors

See Resource Guide for stamp credits.

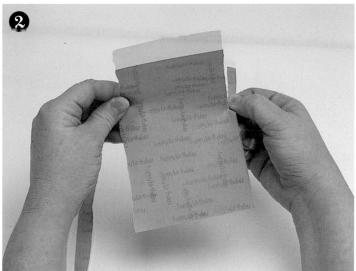

1. Ink and Stamp "Happy Birthday"
Ink the "Happy Birthday" stamp with purple dye ink and stamp randomly onto turquoise Xyroned vellum. (Running vellum through a Xyron machine will add a sticky coating to one side.)

2. Tear Vellum
Tear the edges of the stamped vellum on both sides.

3. Adhere Decorative Paper
Using double-stick tape, adhere a 3" (7.6cm) square piece of decorative paper diagonally onto the front of a folded white greeting card.

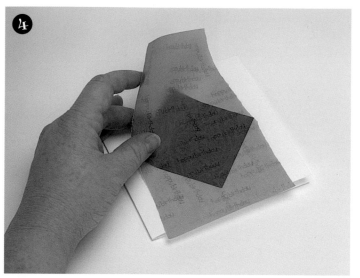

4. Adhere Vellum to Greeting Card
Cover the entire front of the greeting card with the stamped vellum.

5. Ink Cake Image
Ink a birthday cake stamp with silver dye ink.

NOTE: When using a triple stamp pad as shown, ink only one cake image.

6. Stamp Cake Image onto Decorative Paper
Cut out another square of decorative paper measuring 2" (5cm) square. Stamp the cake image in the center of the square.

7. Adhere Decorative Paper to Card
Using double-stick tape, adhere the small square over the vellum on the greeting card, positioning it diagonally over the larger square. Trim the bottom of the card as necessary.

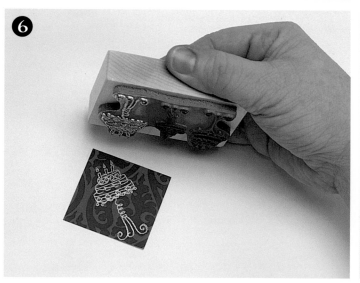

More Bright Ideas

Beautiful Greenery

The layered squares of paper in this card add such elegance, your loved ones will never know how easy it is to make. Choose decorative paper in matching shades, or use the scrap paper you have saved from other projects.

Colors of Fall

Using paper with different patterns and textures will make your cards more interesting and fun. Vary the shapes of the paper as well. Here, the circle among all the squares really sets off the simple leaf.

COMFY AND COZY

masking can be so impressive when done well—and Eclipse tape makes it super simple. Look for stamps with open areas, like this overstuffed couch, so you can use all your favorite small designs to decorate the fabric.

TIP

Creating a plaid pattern with a word stamp is easy. Stamp the words so they run both horizontally and vertically and use varying amounts of ink on the stamp to create light and dark lines.

What you'll Need

Stamps*

- Comfy chair stamp
- "Thank You" stamp

Materials

- Folded white greeting card
- Dye ink, various bright colors
- Black solvent ink
- Orange paint pen
- Eclipse tape
- Craft knife

See Resource Guide for stamp credits.

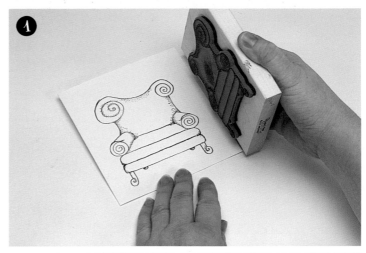

1. Ink and Stamp Chair Image

Ink the comfy chair stamp with black solvent ink and stamp onto a folded white greeting card.

2. Roll Eclipse Tape Over Stamp

Ink the stamp again and roll Eclipse tape over it to transfer the image.

3. Cut Out Chair Image

Cut out the chair image from the Eclipse tape with a craft knife. You will be left with an outline of the chair, which will be your mortise mask. (For more information on masking, see page 19.)

NOTE: Save the chair cut-out to use as a regular mask on another card.

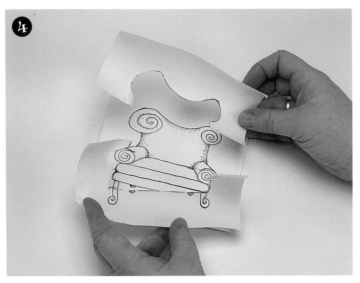

4. Cut Mask in Two Pieces
Cut the mortise mask into two pieces. This will make it easier to work with.

5. Apply Mask and Stamp "Thank You"
Apply the mask around the chair image on the greeting card. Ink the "Thank You" stamp with various colors of dye ink and stamp it horizontally and vertically onto the greeting card. Remove the mask.

6. Mask Inside of Greeting Card
Cut out another piece of Eclipse tape and stick it to the inside of the greeting card, masking all but the far right edge.

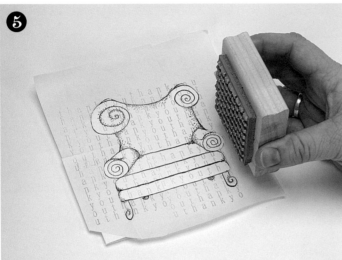

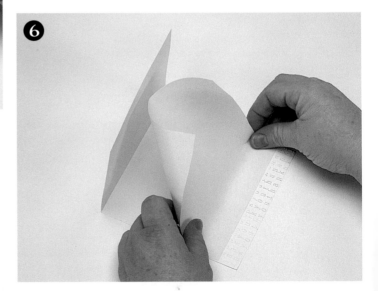

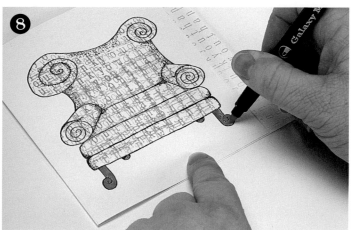

7. Ink and Stamp "Thank You"

Ink the "Thank You" stamp with various colors of dye ink and stamp the inside edge of the greeting card. Remove the mask and trim the front of the greeting card along the right-hand side so that the inside edge is visable.

8. Color Chair Feet

Color the feet of the chair with an orange paint pen.

MORE BRIGHT IDEAS

Have a Seat

Use the regular and mortise chair masks to create cards with different patterns and colors on the chair and walls. If you plan to apply the masks over and over, I suggest using a heavier paper, as it will hold up to more wear and tear.

COCKTAILS, ANYONE?

dusting metal pigments over wet pigment ink can give a card an elegant, matte metal look. When combined with an array of whimsical stamps, this card is the perfect mix of sophistication and fun. Metal pigments are a bit more difficult to find, but well worth the effort. They truly look like foil leaf and are usable with any type of pigment ink.

TIP

Keep the metal dust under control with a lint roller. Simply roll over the metallic area of the finished card to pick up any remaining dust. I use lint rollers for cleaning up glitter and other dry pigments as well.

What you'll Need

Stamps*

- Cocktails stamp
- Filler drink stamp

Materials

- Green cardstock
- Silver cardstock
- Folded silver greeting card
- Black pigment ink
- Gold pigment powder
- Spray fixative
- Paper cord
- Soft brush
- Pencil
- Double-stick tape
- Craft scissors

*See Resource Guide for stamp credits.

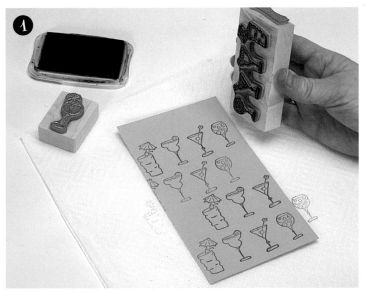

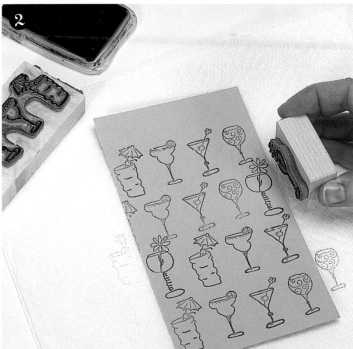

1. Ink and Stamp Cocktails

Ink the cocktails stamp with black pigment ink and stamp it onto green cardstock.

2. Fill in Blank Spaces

Use a filler drink stamp to stamp in the blank spaces.

3. Apply Pigment Powder

Use a soft brush to dust the card with gold pigment powder. The powder will adhere to the wet ink.

4. Remove Excess Powder
Wipe away the excess powder.

5. Apply Fixative
Spray the entire card with fixative (I used Krylon Spray Workable Fixative) to prevent smudging and protect the card from the elements.

6. Adhere Cardstock to Greeting Card
Trim the green cardstock and adhere to the front of a folded silver greeting card with double-stick tape.

7. Stamp and Adhere Image
Stamp one of the images from the cocktails stamp onto a small scrap of green cardstock. Trim around the edges of the image and adhere it to a slightly larger square of silver cardstock with double-stick tape. Trim again as needed.

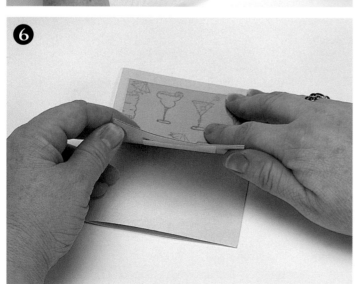

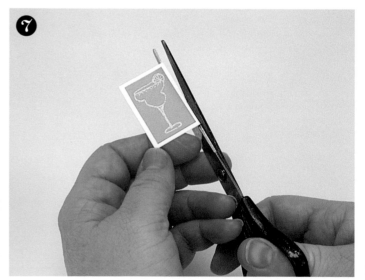

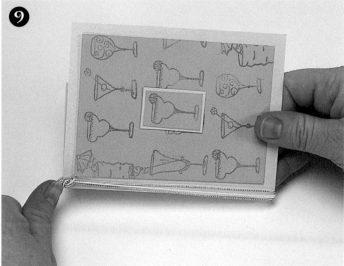

8. Adhere Silver Cardstock
Adhere the small square of silver cardstock to the center of the green cardstock.

9. Add Cord
Add a paper cord to the card and knot at the top.

10. Curl Cord Ends
Curl the ends of the cord with a pencil.

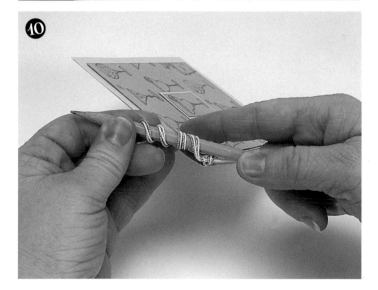

HOME SWEET HOME

When you need to whip up a quick housewarming or hostess card, this is the design for you. Choose stamps with crisp geometric edges to keep this card simple. These house stamps are just the ticket! With a little creative cutting, the colors seem to shine right through the windows.

TIP

To prevent light colors from getting muddy, use the dye stamp pads in order from lightest to darkest.

What you'll Need

Stamps*

• House stamp
• "Party" stamp

Materials

• Folded white greeting card
• Black solvent ink
• Dye ink, various bright colors
• Black permanent marker
• Eclipse tape
• Craft knife
• Ruler

*See Resource Guide for stamp credits.

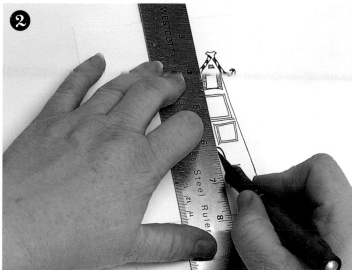

1. Ink and Stamp House Image

Ink the house stamp with black solvent ink and stamp the image onto the right-hand edge of a folded white greeting card.

2. Cut Through House

Open the card and lay it out flat. Using the straight edge of the ruler and a craft knife, cut a border through the house, taking care not to detach the house from the front of the card.

3. Cut Out Windows and Door

Cut out the windows, door and around the right side of the house. The house becomes the right-hand border of the card.

4. Outline Edges

Outline the edges of the house image with black permanent marker.

5. Mask Inside of Card and Stamp with Inkpads

Mask off the inside of the card with Eclipse tape, exposing only the far right edge. Stamp the edge with dye inkpads in varying colors. Remove the mask and allow the ink to dry.

6. Ink and Stamp "Party"

Mask off the front of the card, exposing only the bottom edge. Ink the "Party" stamp with brightly colored dye inks and stamp along the bottom of the card.

7. Add Finishing Touch

To finish the card, use a ruler and black permanent marker to draw a line above the "Party" stamps.

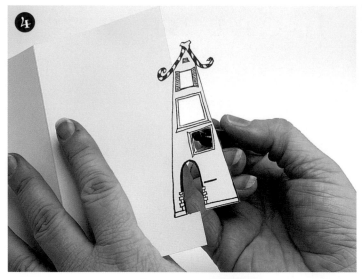

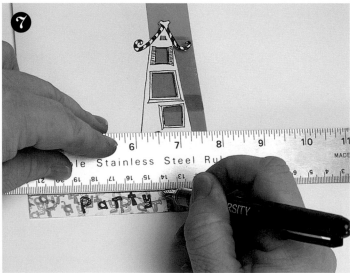

MORE BRIGHT IDEAS

Cut It Out

Stamping the gift image at playful angles makes for a terrific edge for the cut-out. I created the colored background first, then stamped the gift boxes and colored them to complement the background that shows through.

'Tis the Season

Making fancy edges along the side of a card is much easier than it looks. Here, I stamped red and green inkpads and silver stars on the inside edge and colored in the house stamp with markers.

Resources

Stamp Suppliers

American Art Stamp
3870 Del Amo Blvd.
Suite 501
Torrance, CA 90503
Phone: (310) 371-6593
Fax: (310) 371-5545
• Stamps used: Present; Little gift; "You're Invited"; Triple hearts; Lamps; Chinese lantern; Dress; "Happy Valentine's Day"; Birthday cake; "Happy Birthday"; Cocktails; "Party"

Art Gone Wild
3110 Payne Ave.
Cleveland, OH 44114
Phone: (800) 945-3980
Fax: (888) 401-2979
www.agwstamps.com
• Stamps used: Box template

DeNami Design
P.O. Box 5617
Kent, WA 98064
Phone: (253) 437-1626
Fax: (253) 437-1627
www.denamidesign.com
• Stamps used: Baby bottle

Earth Tone Images
P.O.Box 814
Havertown, PA 19083
Phone: (610) 645-6500
Fax: (610) 645-9631
www.earthtoneimages.com
• Stamps used: Handprint

JudiKins
17832 S. Hobart Blvd.
Gardena, CA 90248
Phone: (310) 515-1115
Fax: (310) 323-6619
• Stamps used: Leaves and butterfly; Swirl: Funky swirl; Fall leaves; Star; Harlequin background; Ballerina; Nile flowers stained glass; Primitive flower; Minstrel; Retro circle; Mini squares; Polka dots

paula best and co.
507 Trail Dr.
Moss Landing, CA 95039
Phone: (831) 632-0587
www.paulabest.com
• Stamps used: Menorah

Postmodern Design
P.O. Box 720416
Norman, OK 73070
Postmoderndesign@aol.com
• Stamps used: David; Boy and girl; Molly's face

Post Script Studio/Carmen's Veranda
P.O. Box 1539
Placentia, CA 92871
Fax: (714) 528-4529
www.postscriptstudio.com
postscriptstudio@msn.com
• Stamps used: Shrine; Marshmallow bunny

River City Rubber Works
5555 S. Meridian
Wichita, KS 67217
Phone: (877) 735-2276
Fax: (316) 529-8940
www.rivercityrubberworks.com
• Stamps used: Leaf

Rubbermoon
P.O. Box 3258
Hayden Lake, ID 83835
www.rubbermoon.com
rubbermoon@nidlink.com
• Stamps used: Pumpkin head; Pumpkins; House; Christmas house; Little bird

Stamper's Anonymous
Williamsburg Square
25967 Detroit Rd.
Cleveland, OH 44145
Phone: (440) 250-9112
Fax: (440) 250-9117
www.stampersanonymous.com
• Stamps used: Large frame stamp; Small frame stamp

Stampotique
9822 North 7th St.
Suite 7
Phoenix, AZ 85020
Phone: (602) 862-0237
Fax: (602) 862-0238
www.stampotique.com
• Stamps used: Little big head; Comfy chair; "Thank You"

Other Suppliers

American Crafts
165 N. 1330 W. B3
Orem, UT 84057
Phone: (800) 879-5185
Fax: (801) 226-5086
www.americancrafts.com
• Galaxy markers and gel pens

Amy's Magic Leaf
173 Main St.
West Leechburg, PA 15656
Phone: (724) 845-1748
• Paper Foil Ephemera

Coffee Break Design
P.O. Box 34281
Indianapolis, IN 46234
Fax: (800) 229-1824
• Eyelets

Colorbox
P.O. Box 98
Anacortes, WA 98821
Phone: (800) 448-4862
www.clearsnap.com
• Inks

Envelopes Please
9685 Kenwood Rd
Cincinnati OH 45242
Phone: (513) 793-4558
www.stampawayusa.com
• Templates

Golden Paints
188 Bell Road
New Berlin, NY 13411
Phone: (607) 847-6154
www.goldenpaints.com
• Acrylic paints

Jacquard
540 Lake Cook Rd.
Suite 160
Deerfield, IL 60015
Phone: (800) 442-0455
Fax: (847) 945-8704
www.jacquard.com
e-mail: jacquard@jacquard.com
• PearlEx powdered pigments

Magenta
2275 Bombardier
Sainte-Julie
QC Canada J3E 2J9
Phone: (450) 922-5253
www.magentastyle.com
• Pewter stickers, printed papers

Marvy Uchida
Phone: (800) 541-5877
www.uchida.com
• Inks

On the Surface
P.O. Box 8026
Wilmette, IL 60091
Phone: (847)675-2521
• Threads and fibers

Really Reasonable Ribbon
P.O. Box 199
Sugar Loaf, NY 10981
Phone: (845) 469-3821
www.reasonableribbon.com
• Ribbons

Speedball Art Products Company
2226 Speedball Rd.
Statesville, NC 28677
Phone: (800) 898-7224
Fax: (704) 838-1472
www.speedballart.com
• Carving supplies

Suzi Finer Artworks & Artware
238 S. Robertson Blvd
Beverly Hills, CA 90211
Phone: (310) 360-1800
Fax: (310) 360-1801
www.suzifiner.com
e-mail: suzi@suzifiner.com
• Pewter stickers

Tsukineko, Inc
17640 N.E. 65th St.
Redmond, WA 98052
Phone: (800) 769-6633
Fax: (425) 883-7418
www.tsukineko.com
e-mail: sales@tsukineko.com
• Inks

USArtquest
7800 Ann Arbor
Grass Lake, MI 49240
Phone: (517) 522-6225
www.usartquest.com
• Pinata inks

Index

Index

GET CREATIVE WITH NORTH LIGHT BOOKS

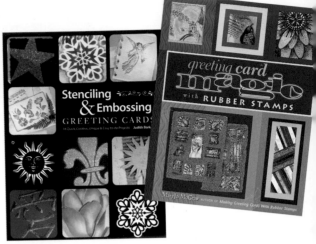

The Essential Guide to Handmade Books

Gabrielle Fox teaches you how to create your own handmade books—one-of-a-kind art pieces that go beyond the standard definition of what a "book" can be. You'll find 11 projects inside. Each one builds upon the next, just as your skills increase. This beginner-friendly progression ensures that you're well prepared to experiment, play and design your own unique handmade books.

ISBN 1-58180-019-3, paperback, 128 pages, #31652-K

The Big Book of Greeting Cards

This book presents a variety of fun, festive and stylish ideas for making cards perfect for any occasion. Discover more than 40 step-by-step projects using a wide range of techniques including rubber stamping, stenciling, quilling and embroidery.

ISBN 1-58180-323-0, paperback, 144 pages, #32287-K

How to Be Creative if You Never Thought You Could

Let Tera Leigh act as your personal craft guide and motivator. She'll help you discover just how creative you really are. You'll explore eight exciting crafts through 16 fun, fabulous projects, including rubber stamping, book-making, papermaking, collage, decorative painting and more. Tera prefaces each new activity with insightful essays and encouraging advice.

ISBN 1-58180-293-5, paperback, 128 pages, #32170-K

Stenciling & Embossing Greeting Cards

Judy Barker introduces you to the basics of stenciling and embossing attractive greeting cards. You'll also learn how to embellish them with foil, polymer clay, shrink plastic and more. It's everything you need to make one-of-a-kind cards for family and friends alike.

ISBN 0-89134-997-9, paperback, 128 pages, #31613-K

Greeting Card Magic with Rubber Stamps

Discover great new tricks for creating extra-special greeting cards! Pick up your stamp, follow along with the illustrated, step-by-step instructions inside, and ta da! You'll amaze everyone (including yourself!) with your beautiful and original creations.

ISBN 0-89134-979-0, paperback, 128 pages, #31521-K

These and other fine North Light titles are available from your local art & craft retailer, bookstore, online supplier or by calling 1-800-448-0915.